Sticks & Stones

Sticks & Stones

A HOLLIS MORGAN MYSTERY

R. FRANKLIN JAMES

Seattle, WA

CAMEL
PRESS

Camel Press
PO Box 70515
Seattle, WA 98127

For more information go to: www.camelpress.com
www.rfranklinjames.com

Cover design by Sabrina Sun

Sticks & Stones
Copyright © 2014 by R. Franklin James

ISBN: 978-1-60381-919-0 (Trade Paper)
ISBN: 978-1-60381-920-6 (eBook)

Library of Congress Control Number: 2014933172

Printed in the United States of America

Praise for R. Franklin James

The Fallen Angels Book Club

"Twists and turns keep this debut novel exciting to the surprising end."

—Michele Drier, author of *Edited for Death* and *The Kandesky Vampire Chronicles*

"Hollis is a character you sorta warm up to. You have to get past her cool exterior and suddenly you realize you REALLY like her and care what happens to her. I ended up finishing the book in one sitting! Although the book is set in the Bay Area outside San Francisco, the author serves up the area as a backdrop, but then concentrates on the characters. The place never intrudes on the story line, like it can in some books. The author does an excellent job of building up the different characters, so that you feel like you know them all. I was happy with the ending (and no I'm not giving it away!) and I'm thrilled there is already another Hollis book in the making!"

—Bless Their Hearts Mom Blog

"R. Franklin James' new book has everything a reader could ask for in a good mystery: intriguing plot, fascinating characters, and a few shockers thrown in along the way."

—Shirley Kennedy, romance novelist

5 Stars: "This mystery kept my attention from beginning to end. Although I had my suspicions of one character, the solution to the mystery surprised me. And that, my friends, is the mark of a good mystery."

—Self-Taught Cook Blog

"A fast paced plot with many twists coupled with a smart and determined protagonist make this a most enjoyable read."
—Kathleen Delaney, author of the Ellen McKenzie real estate mysteries

"A smooth running story where slowly pieces of the puzzle are revealed. Being a book lover I liked the setting she created with the book club …. The author manages to reel the reader in with her delightful storytelling and likable characters. It's a great first book that lovers of the old-fashioned detective genre surely will appreciate!"
—Fenny, Hotchpotch Blog

"A satisfying, clean mystery with several twists that kept me guessing, and also left me anxious for the next book in the Hollis Morgan Mystery series."
—W.V. Stitcher

"A fast paced mystery that keeps the reader wanting more. I love a good mystery and this is one of the better ones I have read in awhile. A fun story for sure!!!"
—Kathleen Kelly, Celtic Lady's Reviews

"The story line was interesting, as were the characters. The book flows steady without any dull gaps. I really liked the author's way of writing…. If you love a good murder mystery, you should get a copy of this book."
—Vicki, I'd Rather Be at the Beach Blog

"This debut novel set in the San Francisco Bay area is a first-person account told by amateur sleuth, Hollis Morgan. This is a very intriguing murder mystery with a startling cast of characters. This book allows the reader to take part in the investigation; I felt my suspicions sift as each new clue was revealed. This is a remarkable, well-rounded mystery and I

HIGHLY recommend this to anyone who enjoys crime fiction."
—Heather Coulter, Books, Books, and More Books

"A very fast and enjoyable read and a real page turner. I was constantly wanting to know what was going to happen next and it managed to keep me guessing a lot of the time as well …. If you love a mystery novel that doesn't bog you down with unnecessary details but still has enough of a background and storyline to keep everything flowing, then I highly recommend giving this one a try."
—Beagle Book Space

"This first book written by Ms. James is a winner for anyone who enjoys a clean mystery which will keep you guessing until the end about 'whodunit.' "
—My Home of Books Blog

"This book is full of murder, mystery and of course mayhem. Thoroughly entertaining and a fast read, I can't wait for the next book in the series. Excellent debut novel, Ms. James!"
—Tammy & Michelle, Nook Users' Book Club

"This is R. Franklin James' debut novel, a fact which I find hard to believe. She has created a character I love in Hollis Morgan, and a great plot …. I'm going to follow the series and R. Franklin James. I've found a winner."
—Views from the Countryside Blog

"Highly inventive … a wonderful thriller. The tension mounts as Hollis becomes the target of the killer putting her life in great peril. *The Fallen Angels Book Club* is loaded with twists and turns and red herrings that will leave you guessing all the while you are flipping pages to find out what happens next. When you finish this book you will heave a hugely satisfying sigh because you have enjoyed yourself immensely.

Ms. Franklin James has provided us with a great new character in Hollis Morgan. I am already looking forward to the next book in this series from this very talented author."
—Vic's Media Room

"The author … does a excellent job of creating a story line that's both realistic and suspenseful. There was never a dull moment. I really look forward to reading more from this author."
—Heather, Saving for 6 Blogspot

"A delightful read. It certainly contained mystery, murder and mayhem …. Like any good mystery, there was a mystery within a mystery and I found [Hollis'] exchanges with the older folks at the center refreshing and decidedly touching …. The reader could feel Hollis's fear with each event and her determination to clear her name. Very well written and very well thought out! Well done Ms James, well done!"
—Beth, Art From the Heart Blog

"*The Fallen Angels Book Club* by author R. Franklin James is an enjoyable first book in the new series featuring Hollis Morgan. Hollis is a good heroine as she is smart, determined and resourceful."
—Barbara Cothern, Portland Book Review

This book is dedicated to Renard Franklin, Mary Sidney, Michael Shannon, and Patsy Williams Baysmore, thank you for being in my life.

ACKNOWLEDGMENTS

IT IS RARE to find a critique group whose professionalism, talent, and caring can elevate a work-in-progress from a rough manuscript to a work of fiction. I have found that group in Kathleen Asay, Patricia Foulk, Terri Judd, and Cindy Sample.

Thanks to Linda Townsdin and Michele Drier, good writers in their own right, who never seem too tired to read my final drafts.

Special thanks to Vanessa Aquino, Barbara Lawrence, Joyce Pope, Geri Nibbs, and especially Leonard James for their much appreciated support.

To my agent, Dawn Dowdle, whose patience is legendary.

Finally, to Catherine Treadgold and Jennifer McCord, my publishers and editors who never cease to amaze me with their ability to find the right words.

Sticks and stones may break my bones … but if you want to hurt someone way down deep, use words.

—Charles Martin, author

Sticks and stones may break your bones and words may break your heart.

—R. Franklin James

Also by the Author:

The Fallen Angels Book Club

CHAPTER ONE

———⁓———

IT NEVER CEASED to amaze Hollis how normal things could look on the outside when all hell was breaking loose on the inside. After taking a last look at the sailboats passing at a leisurely pace under the San Francisco Bay Bridge, she closed the window shutters. With renewed determination, she turned to tackle the maze of boxes in the room behind her.

Packing tape bulged along the edges of the storage box with a finality that marked not only the end of Hollis' law studies but the conclusion of a turbulent chapter in her life. The box was too heavy to lift, so she kicked and pushed it down the hallway to her spare bedroom, where it joined the growing stack of boxes, old clothes, and items waiting for a final resting place. Shutting the door firmly, she leaned against the door frame and caught her reflection in the hallway mirror. Her thick auburn hair had slipped from the bun atop her head, and a pair of tired brown eyes gazed back at her.

The California Bar was done and so was she. Law schools wanted students with life experience. But she doubted it included thirty-three-year-olds who had taken time out for a stint in prison. That sad chapter in her life began with a conviction for a white-collar crime orchestrated by her now deceased ex-husband. Fortunately the prison sentence that followed—eighteen months of real incarceration time—was not the end of her dreams. She fought to clear her name and received a rare pardon. Since then, Hollis tried to make up for lost time by making each day count.

She turned off the light and headed downstairs. It had been a long road and now she was ready to hit the restart button. After four months she was returning to her old job at Triple D, or as the letterhead put it, "The Law Firm of Dodson, Dodson and Doyle."

She sprawled out on the couch and closed her eyes. Just a few minutes of pause; then she'd finish returning everything to its place.

The ringing of the kitchen phone jarred her awake.

Adrenaline pumping, she squinted up at the clock and went to the kitchen.

"Yes?"

"Hollis, this is Cathy Briscoe. I'm sorry for calling so late. But I need to talk to you. Can I come over?"

"Cathy? What's wrong?" She opened her eyes wide. "It's almost eleven."

"I'll tell you when I get there. Still living in San Lucian, right?"

"Yes, but here—"

"Hollis, one day you need to look at why you don't like friends to visit your home. But no matter, I don't have time to deal with your phobias."

"It's a preference, not a phobia." Hollis sighed. "No, I haven't moved. Sure, come over, but I can't promise I won't fall asleep on you."

"I won't take long, I promise. I'll be there in ten minutes."

Catherine Briscoe used to work for Triple D. She had gone to Hastings Law School and shared her first year with Hollis, but unlike Hollis she had finished on time. She took the bar and returned to Triple D as one of their top notch associates. While Hollis specialized in getting through her sentence for insurance fraud in Chowchilla Prison, Cathy specialized in intellectual property and fit right into Triple D's boutique law niche. When Hollis followed in her footsteps and got hired by Triple D as a paralegal, Cathy took her under her wing and wouldn't let her get too discouraged or too vengeful.

Hollis straightened the kitchen table and chairs. She turned on the living room light, put water on for tea, and placed two wine glasses in the freezer.

No one at Triple D was surprised when, after two years, Cathy was lured away by McClouds, one of the largest international law firms in San Francisco, but she never forgot Hollis. Over the last couple of years they had kept in contact, meeting for drinks or meals on a regular basis. In recent months, while Hollis was studying for the bar, phone conversations and emails with Cathy had taken the place of visits. Cathy had offered to play proctor for Hollis' practice exams. Hollis appreciated the offer but declined; she did her studying better alone.

Ten minutes later she heard Cathy's knock on the front door. Cathy, at five four, was only a little taller than Hollis. Her hazel eyes were framed with thick dark eyebrows and her long dark hair was pulled back into her trademark braid. She carried a briefcase, tote, and manila folder stuffed with papers. The minute she was in the door, she had her jacket half off. Hollis couldn't help but notice the sheen of sweat on her cheeks and forehead.

After a few moments of catching up, they settled onto the living room sofa and Hollis set two cups of tea on the coffee table.

Cathy looked around. "Have you got any wine?"

"Sure." Hollis got up. "Glasses are in the freezer. Start talking; I can hear you."

"Thanks, Hollis. I'm sorry to bother you. I remember what a mound of babble I was after I took the bar, but you're the only one I could think of who might be able to help me. Mark thought so, too."

Mark Haddan was an attorney who used to work at Triple D. After he also moved on to McClouds, he and Hollis remained good friends. She owed him.

"Mark is pretty savvy," Hollis called out from the kitchen. "Go ahead, let's hear it."

"I didn't tell you because you were deep into studying, but I left McClouds about five months ago. I'm now a freelance writer covering stories with a legal slant for *Transformation*. You know the one that greets you at the check-out in the grocery store?"

Hollis raised her eyebrows but said nothing. *Transformation* was cousin to *Intercept* and *Speakeasy* tabloids.

"Why didn't you tell me the last time we talked on the phone?" Hollis put a tray with glasses and a bottle of white wine on the table.

Cathy shook her head and hugged herself, gripping her upper arms. "I thought about it, but you were finishing up at Hastings and then studying for the bar. Anyway, I was burning out on the law, and I realized I liked taking on freelance assignments." She paused. "Last month I submitted a story about a non-profit group with ties to Dorian Fields and his Fields of Giving organization." She took a sip of wine. "I cleared all the research and we had a ton of corroborating testimony."

"Sounds good."

She nodded. "It was good. It was my first investigative article. Dorian Fields is masquerading as a warm-hearted entrepreneur while using his non-profits as shells for money laundering. I got photos and tapes. Here's a copy of my article. It was to be the first in a three part serial."

"Wow, I get it." Hollis glanced quickly over the pages. "Where did you get your hands on evidence like this?"

Cathy raised her glass in a mock toast. "A whistleblower who wanted to remain anonymous. It's not uncommon with tabloids. Anyway I went through each piece of information. It all checked out."

"Okay. Do I hear a Pulitzer?"

She snorted. "Last Wednesday, I was served a lawsuit for libel. Mr. Fields asserts my story was completely bogus. He's suing me for twenty million."

Hollis paused with her pen in the air. "You, personally, not the paper?"

Cathy nodded. "Both. *Transformation* has me under a freelance contract. I'm still on probation. I knew it was risky, but I couldn't imagine anything would happen that would put me at liability. I've practiced intellectual property law; I made sure I vetted copies of the evidence with our attorneys …." Her voice trailed off.

"Did it actually get published?"

"No … well yes, *Transformation* distributes copies on Thursdays. This issue was stopped at the warehouse. Boy, was management pissed."

"How did Fields find out?"

"I have no idea. Although, in a way, it saved me. If the paper had been distributed, the suit would have been three times as large."

Hollis looked up from the pad in her lap. "What can I do? I'm just a law student anxiously waiting for the next months to go by quickly so I can know by Thanksgiving if I passed the bar exam."

"While I was lamenting to Mark, he reminded me what a great researcher you are. Look, Hollis, I don't have a lot of money. I'm still paying off law school loans. I need you to help me re-gather the information on Fields. It took me three months of twenty-four/seven to pull together everything the first time. I don't have that leeway anymore. My hearing is next month. I already asked for an extension. I'm *pro per*."

"Are you sure you want to represent yourself? Wait a minute, I'm confused. Why do you need research? You already found all the information to back up the article."

Cathy balled up her fists and hit the arms of her chair. "Two nights ago all my research was stolen. I had to go down to L.A. to take care of a financial matter, and when I got home the house was a mess and my research papers were gone. I had copies on my laptop and a backup on my flash drive. They got both."

"Did you call the police?"

"Yes, of course. But I was also missing my TV, my music system, and some jewelry. So they said it was a routine burglary. There was nothing that pointed to Fields."

"So, you think he was getting rid of your proof before the court hearing?"

"You better believe it."

"But why me? Surely, you know people who are better at this than I am."

Hollis couldn't shake the feeling there was more to her friend's story than she was saying.

Cathy nodded with understanding.

"I remember when your doggedness solved the murder and fraud case at Triple D. I was at McClouds, but everybody there knew you saved Triple D a ton of money. The number of lawsuits was minimal, because Triple D got jump on the client settlements."

"But that wasn't libel."

"Hollis, what's the only real defense against libel?"

"The truth."

"Exactly. Dorian Fields knows that I know the truth. He has to stop me from ruining his name."

Hollis didn't want to point out he was off to a good start. She ran her fingers through her hair.

"Okay, okay. Look, it's almost midnight and I'm beat. I return to the office tomorrow for the first time since I went on leave to study for the bar. I can't think straight. Let me get some rest. I'm not a night person and I need to think through what you've told me. I'll call you tomorrow."

"No, you don't understand. I don't mean to be unreasonable, but I haven't slept in days. I need to know now that you'll help me. Just take a look at this." She reached into her tote and brought up a file filled with clippings.

Hollis patted the file without holding it. "Cathy, I can't. Give me until tomorrow morning. I'll call you first thing—promise. My brain feels like oatmeal. I want to help you, but if I had a little sleep, I might even be able to think of somebody who could be more help."

Cathy's shoulders slumped and she closed her eyes. When she opened them, Hollis could see they were bloodshot and glistening.

Resigned, she gathered her tote and briefcase. "Tomorrow then."

BEFORE LEAVING FOR work the next morning, Hollis glanced through the material Cathy left behind and made up her mind to help her. It shouldn't take too long to substantiate the assertions, and she already had a couple of ideas—first, she would interview the nonprofits. Punching Cathy's number in her cellphone she was greeted with a monotone response. Hollis left a voice message to get in contact.

TRIPLE D'S LOBBY hadn't changed. Other than having the name of one of its partners removed from the directory, it was the same as it had been for years. Hollis thought it would look or feel different to her after being on leave for the past three months. It didn't. Taking advantage of Triple D's employee professional education assistance program, she'd converted her vacation time to paid hours that were matched by the firm.

"Welcome back!" Doe-eyed Tiffany smiled up from her perch at the receptionist desk. Her perky wholesome looks masked a sharp mind. "How did it go?"

Hollis took a deep breath. "It was grueling. I don't know if I did well or not, I'm just glad it's over. I've started to—"

"Hollis, good morning, you're here bright and early, just like the old days." Ed Simmons, the firm's managing partner, came through the lobby doors with briefcase and iPhone in hand. "We really missed you. There are a couple of probate cases that are giving George grief. I'm sure he'll be even happier to see you than I am."

Without waiting for her answer, he walked past into the hallway leading to his office. Ed's appearance was as no-nonsense as his manner. Plain suit and tie, nothing flashy, and a face few witnesses would be able to describe because of its ordinariness. His receding brown hair, capped regular features, and with eyes a little too close together, very little escaped him.

Tiffany made a sympathetic face and picked up the ringing phone.

Hollis grinned and shook her head. Ed had not changed either; he was still in charge and the firm would always come first. It was good to be back.

Triple D had been her refuge in the past and she hoped it would continue to be in the future. The last months had been a mesh of study, intense study, and mind-bending study. Preparing for the bar exam had consumed her life. At the end, after hundreds of hours, it came down to three days, two performance tests, six essays and two hundred multiple choice questions. She wouldn't find out the results until November, three more months, which was fine with her. It would take her that long to recuperate. Now, back at work, she found her office to be both familiar and distant at the same time, like looking at old photographs of herself as a child. No one had moved a thing since she left a month ago.

"Good morning. Am I glad to see you!"

Hollis smiled at the tall man in her doorway. George Ravel was a good boss. He had twenty years on her in age, but his youthful outlook and good nature closed a good part of the gap. He dressed like he lived, down to earth. She was glad he hadn't changed. He was more like a peer.

"You know, it feels great to be back."

"How did it go?" He occupied the only guest chair in her office.

Hollis took the next few minutes to recount bar exam details. George listened intently. He could appreciate her description and knew exactly what she had been through.

"Anyway," Hollis concluded, "except for the last essay I'm feeling pretty good. But I've heard much the same from people who've crashed and burned."

"Hey, hey, don't look so glum. It's all over now."

"George, almost half the room was filled with people who had failed the exam at least once. One guy failed three times."

"What's your point?" George said. "They're not you. But now that you bring it up, I kind of like the idea of you being my paralegal for life."

Hollis laughed. "No way, buddy, take advantage of my skills now 'cause come fall, I'm out of here."

He grinned. "Good, that sounds more like you. But while you're still here I could sure use your help. Take some time to get back into gear. Then let's meet around ten o'clock to go over cases with our new associate."

Hollis glanced over to her in-box.

George had meant what he said; he was glad to see her. She glanced through the stack of filings and a shorter stack of correspondence needing responses. After thirty minutes of flipping through files, she realized that—improbable as it was—some of these cases seemed to be where she'd left them months ago. She felt challenged, rather than overwhelmed, knowing she'd have everything up to date in short order.

Hollis punched in Cathy's cell number and got her machine again. "Cathy, you've sold me. I want to help. Give me a call at the office." She also sent a text message.

Where was Cathy? Last night she'd pushed hard for an instant decision.

In George's ten o'clock meeting, she met Tim Walker, a new associate attorney on their team. Hollis could tell this was his first job. Young and a little gawky, he was rightly embarrassed when Ed, on his way out, pointed to his one navy blue and one black sock. In response, the new associate's cheeks and rather prominent ears turned beet red. Hollis, concentrating on going through papers, pretended not to hear or see a thing.

George followed her example.

"Hollis, we have a case that came in right after you left. It took a while to ascertain there were no heirs, and now the client's house is listed for sale. The furniture needs to be inventoried and sold through an auction house. Ordinarily, I'd give it to you, but …." his voice uncharacteristically drifted.

"I'll be fine. I'm not traumatized." She reassured him. She knew George hesitated because the last time she inventoried a client's assets it resulted in her being left for dead. She had trusted her then-manager emotionally and professionally and he had joined her list of betrayers. But other than random unintentional looking back—she didn't. She refused to be defined by obstacles.

"Good." George read from a thin folder. "Margaret Koch was a Triple D client who died and left the firm the executor of her estate. No relatives came forward, and because of the size of her estate, we hired an investigator to make sure there were no heirs."

"If Hollis is handling Koch, I'll file the request for the Miller probate hearing," Tim said, scribbling like a maniac. "Uh, what court are we talking about?"

"Probate Court, Department Seven." George looked at him with curiosity. "If you need help, let Hollis know. We're a little behind on this one but it's not complicated."

Hollis reached for the file. "Sure, be glad to help."

Pulling back, Tim fumbled and dropped his papers on the floor then stooped to gather them up. "No, no, it's okay. I got them. I'll get these back in order, Department Seven, no problem."

George and Hollis exchanged doubtful looks.

George nodded. "Okay, Hollis, you conduct the inventory. Let's try to get everything wrapped up by the end of the week."

HOLLIS LAID THE armful of file folders on her desk and put the envelope that held keys to the Koch house in her purse. Hearing her stomach give a low growl, she decided to stop for a quick bite of lunch then head out to the property. She wanted to get some idea of the task ahead. She had until the end of the week, but she couldn't stop being compulsive—no, she didn't want to stop. She liked being compulsive. If she gave it some thought, the trait probably stemmed from her need to control situations; much of her early life had been out of her control.

She tried to reach Cathy again, but still there was no answer. If Hollis didn't catch up to her by the time she finished at the Koch house, she would go by her place. She just hoped Cathy hadn't given up on her.

THE KOCH HOUSE had major creaks.

There was another one.

Hollis tilted her head toward the unseen noise. She didn't frighten easily but the old house, with its dark corners and creaking walls, tested her resolve. After ten minutes, she began to regret not bringing one of the firm's junior paralegals with her to help take photos of the house's contents. She pulled back the heavy drapes. The afternoon was dim with heavy clouds, and the sixty-watt bulbs did the minimum to light the rooms. She paused.

Someone was in the house with her. The sound of movement was fleeting, but not random.

"Hello," she called out.

Footsteps, muffled by carpet, hurried her way.

"Who are you? What are you doing here?" A tall young woman, her sandy blond hair pulled back with a headband, strode into the room.

She looked to be in her twenties, dressed in jeans, a crisp

white blouse and a form-fitting tailored black wool jacket. *Very Town & Country*, Hollis thought, as she brushed off a light film of dust covering her navy blue slacks and tan pullover sweatshirt. She put out her hand. The young woman looked down at it as if deciding whether or not to accept the handshake. She did.

"I'm Hollis Morgan. I work for Dodson, Dodson and Doyle. We represent the estate of the owner, Margaret Koch."

The stranger had a firm handshake and wore a wedding ring.

"Kelly Schaefer. My mother was close to Mrs. Koch. I didn't know anyone would be here. I came to see the old place before it was sold." She turned around and pointed. "We used to celebrate holidays in this room."

"Are you a relative? We've been trying to locate heirs."

Avoiding eye contact, Kelly walked around the room, running her hand lightly over the artifacts and upholstery.

"No. My mom was friends with Mrs. Koch. My mother died years ago. When I read in the paper that Mrs. Koch died I just wanted to check if some of my mom's things got left behind." Kelly put her hands in her pockets.

"Why would Mrs. Koch have anything from your mom after such a long period of time?"

"My dad mentioned that my mom was special to Mrs. Koch. He even brought me by to visit her. Then he died and I was alone." Kelly gave Hollis a rueful smile and moved toward the door. "I just wanted to see it one last time."

She's lying.

Hollis knew when she was being lied to. The instinct was a point of pride and rarely failed her. She only got into trouble when she failed to pay attention.

"Could I have your contact info ... in case I find anything?"

"Sure, do you have a piece of paper?"

Hollis handed over a sheet from her pad. Kelly scribbled her name and number and handed it back. Hollis glanced at it briefly and slid it into her pocket. She wouldn't take odds that the number was any good. She'd first been aware of her knack for sensing lies in childhood. Although it had proved more valuable to her as an adult. Especially in prison, where it had come to her rescue many times.

"Well, I've got a few more rooms to do, and I've got to get going," Hollis said pointedly just as a thought came to her. "How did you get in? The locks were changed."

Kelly looked around the room. "I know. I stumbled on a utility door on the side of the house that opens up to the downstairs. It doesn't have a lock."

It will tomorrow.

Hollis frowned. "Is there something in particular you're trying to find?"

"A box of mementos, nothing to do with anything legal." Kelly scanned the room. "My mother mentioned a token necklace."

"I would appreciate it if you let me see anything you take out of the house." Hollis handed her a business card. "The court has directed us to undertake an accounting of all the assets, even mementos. Things can get kind of crazy if items go missing."

Kelly smoothed her jacket and stared down at her card.

"Sure, sure, I understand."

"Thanks." Hollis looked down at her watch again. "Well, like I said, I've got to finish."

"Oh, don't worry about me, I'm leaving. Do you mind if I go upstairs to use the bathroom?" Kelly motioned with her head.

"There're no working bathrooms on the second floor." Hollis gave Kelly a speculative look.

"Uh, I wasn't going to use it. I left my scarf in the guest bathroom."

She's lying again.

"Sure, okay, not a problem. Let me know before you leave." Hollis pointed to a cabinet. "I'm still finishing up down here. I'll be going upstairs next." She turned back to her camera.

Hollis was still taking photos in the den when she heard

the click of the front door. She walked over to the windows in time to catch Kelly Schaefer walking to a late model Nissan, looking over her shoulder back at the house. Hollis took down the plate number.

She could have sworn that Kelly's stylish jacket showed a bulge that hadn't been there before.

CHAPTER TWO

THE FIRST THING Hollis did when she got back to the office was contact a local locksmith to change the locks on the Koch house. He agreed to get it done no later than the next morning. She was punching Cathy's number in her iPhone just as Tiffany poked her head in the door.

"Hollis, did you hear? Cathy Briscoe was found dead earlier today. The police think it's suicide."

Hollis almost dropped her phone. "Oh my God, no! What happened?"

"I'm not sure. It just came on the news. Everybody is gathering in front of the television in the conference room." She looked over her shoulder. "I've got to go tell the others."

Hollis shivered. With trembling hands, she tried unsuccessfully to pick up the cup on her desk. Leaving it behind, she walked down the hallway toward the blare of the television and took a seat in the far corner.

"Informed sources say the death of Catherine Rose Briscoe appears to be suicide. Ms. Briscoe was a reporter for the tabloid, *Transformation*. According to the statement released by the police, co-workers say Ms. Briscoe may have been distraught after facing a libel suit. Let's hear from the police spokesperson."

"Suicide," one of the attorneys murmured.

"Shhhh, here's the police," another whispered.

"Thank you, at six-fifteen this morning the department's nine-one-one operator was called by a neighbor, who claimed to hear the drone of a running motor coming from a closed garage. The EMTs broke into the unit and discovered the body of Catherine Briscoe sitting in her car and the garage full of fumes. They pronounced Briscoe dead at the scene. Oakland detectives interviewed witnesses, who heard arguing early in the evening. At this time suicide is suspected. We'll know more after an autopsy."

The station cut to a reporter's story on a local coffee house groundbreaking.

"I just can't believe it," Marion Babbitt, the firm's office manager, said. "I just talked to her last week about her 401k. She left it with us, and she wanted to know how to convert it to a Roth IRA. I just can't believe it."

Steve, one of the firm's senior partners, shook his head. "She was a damn good lawyer." He pointed to one of the secretaries. "Find out the particulars of the funeral. Make sure we send a large floral arrangement."

Ed, who had been perched on the edge of the conference table, rose to his feet. "Good idea. I've met her mother. She's a very pleasant lady."

Gradually they all dispersed into the hallway. A low hum of conversation drifted out into the lobby.

Frozen in place, Hollis tried to move, but her legs felt like they were made of stone.

"You okay?" George put a hand on her shoulder. "I only met Cathy once. Actually, it was at her going-away party. She seemed like a nice, smart lady."

Hollis, unable to get any words to form on her lips, simply nodded.

"Were you friends?"

Again a nod.

"Such a shame." George looked down at his watch. "Do you think you'll have that legal brief draft done by five? I don't want to rush you, but if you're upset I can get one of the associates to—"

"I'm not upset." Hollis heard the shaking in her voice. She took a deep breath.

"Hollis, if you're not all right—"

"I'm sorry. No, really, I'm all right. I need to keep working. I've already finished the first cut. You can look it over, make any changes, and I'll finish up before I go home."

"Well, good then, as always you're way ahead of me. You're—"

Hollis put her hand to her forehead. "George, I … I saw Cathy last night."

"What?"

"She … she came over to my house." Hollis swallowed. "She needed … she needed a favor. Maybe I could have saved her."

He frowned. "You've got to tell the police."

Hollis nodded in agreement. This part she knew by heart.

NOT MUCH HAD changed in the months since Hollis had last visited the Oakland Police Administration Building near Jack London Square, although now it appeared to be sharing the sidewalk with some sort of street-art fair. After carefully slipping her car between colorfully dressed pedestrians into a metered parking spot, she found herself just staring out the front windshield. Intellectually she knew she wasn't responsible for Cathy's death, but she couldn't stop thinking that if she had responded to her that night, things could have been different. But even more than that, Hollis wished she had the chance to tell Cathy she wouldn't disappoint her. Maybe she missed the chance to tell her in person, but she'd make sure she'd fulfill her friend's last request.

She stiffened her back and strode into the lobby.

Although she recognized the security guard immediately, she could see him struggle trying to remember who she was. Back then the circumstances had been completely different. She was an ex-felon under suspicion for murder trying to clear her name.

She was glad she'd remembered it was best to call ahead. Drop-in appointments with the Department were too often an experience in waiting with the officer of the day until a detective was free. He checked for her name on a list and waved her past. She went through the metal detector, and retrieved her purse. When the guard started to tell her how to get to her destination, she told him she knew where she was going and walked down the narrow corridor to a small waiting room.

There was a new reception desk manned by a bookish looking officer scribbling notes on a stack of papers.

"Excuse me," she said. "I have a meeting with Detective Faber."

"Have a seat." He looked up briefly. "He'll be out shortly."

Hollis nodded and took a seat in one of the white plastic chairs lining the wall.

Faber came out in less than five minutes.

"Ms. Morgan, I can honestly say I was surprised to get your call." John Faber put out his hand and said in a more subdued voice, "It's really good to see you."

Hollis was caught off guard by his tone. The last time she had spoken with him he was sending her on her way after an admonition to stay out of police business. Then she had left, and gladly. It was only during her hearing for a pardon that she realized he had provided a critical recommendation she needed to convince the judge to give her a second chance.

She stood and smiled. "It's good to see you too, Detective Faber." She was surprised how much she meant it.

He stood aside and ushered her down a long hallway.

Faber turned to her as they walked. "How is everything going? Judge Mathis told me you received your pardon. Have you gone back to law school?"

"Yes, I graduated last spring. In fact I just finished taking the bar."

"I wish you luck." He looked down at her and smiled. "Let's go in here. Under a mutual courtesy policy we get to share a few visitors' spaces with OPD. I've got about thirty minutes. You were a little mysterious on the phone. How can I help you?" They entered a large Plexiglas-enclosed cubicle.

She sat in one of the only two chairs in front of a metal desk. "First of all, call me Hollis. I was glad you could still meet with me. I'm here about a sui … a death reported on the news this morning—Catherine Briscoe?" She still couldn't bear to think that Cathy was desperate enough to kill herself.

"I know of the case. It's not mine, but what do you know about it?"

Hollis smiled to herself as she flashed back to Faber's interrogation skills.

"I know the victim. I saw her last night. She needed my help."

Faber raised his hand, reached for the phone on the desk, and tapped in four numbers. "Cavanaugh—Faber, I've got someone in Office A. You need to talk to her about the Briscoe case. Yeah, we'll meet you over there."

"Let me guess, we're going to the interview room?" Hollis stood.

"Yep, there's more space."

And a two-way mirror with a microphone, thought Hollis.

Cavanaugh was already sitting there when they arrived. Hollis prided herself on being able to size up people fairly quickly. Cavanaugh was intense, with an average build on a paunchy frame. His dark brown eyes pierced through her as he tapped his fingers against the desktop and intermittently ran his hand over an imaginary beard.

His twisted smile seemed to beg her to pick a fight. "I'm Tom Cavanaugh. Thank you for reporting your contact with the victim. Why don't you start from the beginning and tell us what you know about Catherine Briscoe."

Hollis shook off his negative vibes. It didn't take long for her to recount her meeting with Cathy. She noticed Faber checking his watch a couple of times. The thirty minutes must have passed.

Cavanaugh asked, "So, she gave you no indication what form this 'proof' took? No real details?"

"No, we were going to talk about it this morning."

"Where's the file she left with you?"

Hollis reached in and pulled it out of her tote.

"Ms. Morgan, I have to leave now." Faber motioned to Cavanaugh. "I've got to make a call, and I'm late for an appointment. I'll check on the burglary follow-up and see you back at your office."

Cavanaugh nodded as Faber went out the door. He turned to Hollis.

"How did she appear to you?"

Hollis thought back to that point in time. "She was distraught, panicky."

"Panicky?"

"Yes, like she was running out of time," she paused and swallowed. "Like she knew something was going to happen."

"Did she have a portfolio with her?"

"A portfolio?"

Cavanaugh leaned back in his chair. "Yeah, a leather folder with a black cover, a black binding and a royal blue fastener."

Hollis shook her head. "No, nothing like that. She just had her purse and a briefcase." She squinted, trying to remember. "She had a folder with her, the one I just gave you."

"We couldn't find a purse. Did she mention any names?"

"No, just Fields."

"What about a boyfriend? Did she have one?"

"No, I don't think she had a significant one."

Cavanaugh paused and stared hard at her.

Hollis stared back.

He turned away and flipped through the file.

"And this is all she left with you?"

Hollis looked him straight in the eyes. "This is all she left with me."

"Any idea why she might commit suicide?"

"Detective, Cathy was a professional as well as my friend. She wouldn't have taken her own life. She was a fighter. It wasn't in her to give up on something she believed in just because she met with an obstacle."

"Sometimes people change when they're backed into a corner."

"Not Cathy. That's why she came to me, to help her prove the truth." Hollis picked up her things. "I think you need to look deeper. I just don't see her committing suicide."

Cavanaugh looked hard at her for a long minute. Then he slapped his hands on his thighs and stood. "Okay, then. If you think of anything else get in touch with me." He handed her his business card.

Hollis put the card in her pocket. "I let her down. I may be too late to save Cathy, but I can save her name."

"I have to advise you that it is against the law to interfere with an open investigation. Mr. Fields has been extremely helpful to us and if you find out something to the contrary and don't let us know, it's a crime."

Hollis wasn't surprised to hear the official party line. For the first time she wondered who was listening to their conversation.

"I'm very clear about that, I assure you I—"

Cavanaugh's cellphone went off. He looked down at the screen and stood.

"Ms. Morgan, I've got to take this call. Just remember what I said."

CHAPTER THREE

⁂

HOLLIS AGREED TO meet Mark Haddan at their favorite Thai restaurant after work. She noticed that he had become more confident and self-assured in the past months. Wearing an expensive-looking slate gray suit, he also had a new haircut. The style of his dark brown hair flattered his dark blue eyes.

The waiter took their order. Hollis listened as Mark recounted his last conversations with their mutual friend. A story about how Cathy saved him from a client who was giving him grief brought back her own memories of Cathy.

Hollis curled up in the booth with her feet tucked under her. "How's Rena?" She and Rena had been members of the Fallen Angels Book Club. The Fallen Angels weren't an ordinary book club; members had to be white collar ex-felons. Unfortunately, book club members had begun to die in suspicious ways, one by one, which brought a halt to their meetings. Hollis' desperate efforts had turned the police investigation around, and as a result she'd been able to clear her name. Afterwards, the remaining club members tried to stay together but eventually they stopped meeting. Hollis realized that she missed the group.

The waitress came by with a small stainless steel pot and refreshed the white porcelain cups that held their tea.

Mark took a sip. "Rena is good. We are good. She told me to tell you hello."

"Just remember, it was I who introduced you two. I want full credit."

"Doesn't saving your life make us even?"

Hollis smiled and nodded. "Oh yeah, you're right. I forgot. We're even."

They both mustered a weak laugh.

"How did it go with the police earlier?" Mark asked. He reached into his briefcase and laid an almost empty manila folder on the table. It was time to get to work.

Taking his cue, Hollis took out a legal pad and flipped to a page full of notes. "After a lot of procedural rambling by the detective in charge, I learned nothing except that Fields has been 'extremely helpful to the police,' whatever that means."

"What happened when Cathy came to see you? What was her of state of mind?"

"She called me late and asked to come over." Hollis sat up in the seat and leaned forward. "She was agitated, frustrated, and angry. The libel suit was making her crazy. Cathy was convinced someone had stolen her research and had made it look like a burglary. She wanted me to help her with re-gathering the research. She told me that you thought I could help. I told her …." She swallowed. "I told her I would have to think about it."

Mark, who had been taking notes, looked up quickly at the choke in Hollis' voice. "Hollis—"

"Mark, she did *not* commit suicide. The police seem to think she did. I know her, you know her. She never would have taken her own life."

"I agree." He reached over and covered her hand. "I know she didn't."

"I just wish I could have told her I was going to help."

"Don't worry, she knew you well enough to know you'd help her." Mark continued, "Cathy wasn't hasty or impulsive. She had facts to back up her story. Did she give you anything we could use?"

"We? Are you ready to take up Cathy's case?"

Mark gave her a small smile. "I lost a friend, too. That's why I'm here."

Hollis gave him a stack of pages. "I copied the file Cathy gave me before I gave it to Cavanaugh. It's primarily clips from newspaper articles showing Fields getting all these awards from various organizations." She pointed. "Then there are these notes with dates and some with question marks. There are still other notes with the word 'verified' underneath."

Mark glanced through the papers. "There's no way we have time to prep for a trial. We need to put our efforts into the biggest payoff. We need to delay a settlement hearing." Mark pulled out a piece of tablet paper that matched the pad Hollis had in front of her. "The first thing I'm going to do is ask for a continuance. But we have a problem even before that."

Hollis nodded. "I know—standing. The lawsuit was against Cathy, not us. And, she never formally brought us in to work on her case. The court won't recognize us as having a claim. *Transformation* management will look to have the case dismissed or settled, although Fields' attorneys may want the visibility of a trial to punish the magazine."

Mark gave a quick shake of his head at the waitress when she approached the table. Hollis did the same and she turned back to Mark.

"What if we talked to *Transformation*?" She held up her hand to stop Mark's anticipated objections. "We need standing; they have standing. They were counting on Cathy's proof to substantiate her claims. They might give us the resources we need to fight Fields."

"If I represented them, I might advise them against that," Mark replied.

"Believe me; I know they could just let Cathy's reputation swing in the wind, but if we could show them there's a chance to defeat Fields, they might just give us the opportunity to save their butts."

Mark wrote on his pad. "Okay, let's say I'm able to get them to hire us for representation. That's if I can sell the idea to my firm first. Same with you; you're going to need the time from Triple D to do the research."

"Just get the six-month continuance, and I'll get the assignment from Triple D."

"Well, I do have a possible foot in the door. There's a partner in our office who knows Carl Devi, the regional editor for *Transformation,* and its chief administrative officer. There's a good chance we can get a meeting with *Transformation* management."

Hollis smiled broadly. "Mark, that's great." She took out her calendar. "Let me know as soon as you get a day and time. I'll make myself available."

"I thought you'd say yes. I'm shooting for this Thursday." He handed her the folder of papers. "Let's be ready. I'll nail down an appointment and you get started on what's available through public information."

"That sounds like a plan."

Mark said, "I took a quick look. There's not much. But go through everything and sort it out by strength of lead."

"I'm on it."

Hollis reached in her tote and pulled out two slim folders. She'd compiled notes of her own.

Mark pointed to the initials on one of the folders. "You got that from Cathy, too?"

"The notes I've already made are in this folder." She pushed it over to him. "And these are Cathy's, from the file she left behind." Hollis slid it across the table.

Mark put his pen down and flipped through the papers.

"Hollis, tell me these are copies. You gave the originals to the police, right?"

When she shrugged, he groaned.

"Oh hell, no! Let's not start out on the wrong foot." He pointed a finger. "I'm an attorney and you plan to be one. We can both be disbarred."

"Okay, okay I'll make copies and give the originals to Cavanaugh. I had a momentary blackout. It's just with copies you couldn't read clearly what she wrote in the margins. I tried scanning everything, but her scribbles didn't show through." She saw the look of rebuke on his face. "Fine, I'll rewrite her scribbles and take the originals to the police first thing tomorrow."

Mark picked up the file and glanced at the top pages. He frowned. "Fields had her served at home. Here's a letter to Cathy from Personnel. When her bosses at *Transformation* got served, they back-stepped from her so fast they even cancelled her subscription."

"But surely the *Transformation* attorneys vetted her copy. They must be vulnerable to lawsuits on a daily basis." Hollis said.

"Usually, yes, but Cathy was a freelance writer and she could sell her story to anyone she wanted. We need to get a copy of her contract, although there must be a legal disclaimer in every *Transformation* issue. We need to go through it in detail, particularly the limited liability language."

Hollis tilted her head toward Mark. "Take a look inside the folder cover."

Mark reached over and flipped the cover. His frown deepened. "It's facial pencil sketches, dozens of faces." He looked up at Hollis, "Did she mention who they are?"

Hollis shook her head. "I didn't give her a chance to tell me. I didn't notice this until I looked at it before I went to work. And yes, I'll give the original to the police."

"There's no text, no names—nothing." Mark handed it back

to Hollis. "You think there's an index somewhere?"

Hollis stared down at the file and notebook. "She wanted to pique my curiosity so I would help her. Cathy used to draw caricatures of people when we were in meetings. She was actually quite talented. When I worked on her cases she had drawings of clients along the margins of her notes."

"So what are you saying? You think these are sketches of people who could prove her case?"

Hollis nodded. "Yes. Or, maybe they're sketches of the bad guys. Had she lived, she knew this would tempt me. This type of thing makes me crazy with curiosity. Still, she must have thought it could help me help her."

"Or like you said, she just wanted to get you curious. She thought she would be around long enough to explain it to you."

Hollis took a page out of the folder and held it over the light from the small lamp between them. "Nothing."

Mark shook his head. "You've got to be kidding me. Did you think you would find handwriting impressions? You've been watching too much television."

"I haven't had time to watch television, mystery or otherwise. I was a law student."

Mark riffled through the pages. "I know Cathy. She would fight to keep her reputation from being tarnished by a crook like Fields." He began placing files back in his briefcase. "I'm going to follow through with *Transformation*. And I'm going to the police and tell them the same thing you did—she'd never have committed suicide." Mark paid the check and they both rose from the table.

Hollis squared her shoulders. "I'm looking forward to working with you on this. I'm pretty sure George will lend me out. Then, once we convince this Carl Devi that we can save his tabloid some money, we can get together and divide up the work."

"Rena and I want to take you out to dinner." Mark held the door open for her. "She'll kill me if I don't bring her in on this."

"Well, that's one thing I got right," Hollis said with only a touch of humor. "Tell her to try not to look like a New York model. My ego is too fragile."

"ARE YOU SURE you can work the Briscoe matter and keep up your work?" George was leaning in her office doorway. "This isn't pro bono, is it? I can spare you for a few weeks but you'll still need to work on a few cases, and you'll have limited overtime."

Hollis stood and closed the folder sitting on her desk. "Mark Haddan is negotiating the terms of our work agreement with *Transformation*. And I will absolutely stay on top of your caseload."

"Which reminds me, were you able to arrange for the auction house to take in the Koch inventory after the movers?"

"Yes. By the way, when I was taking inventory, there was this young woman who was wandering around the Koch house. She said she and her mother used to visit. Any risk she could be a relative?"

"Not a chance."

"Still, I'm going to track down her license plate. After trying to reach her I realized she lied to me about her real phone number."

His eyes narrowed. "Interesting. Why would she lie about that?"

"Do you think she could she be setting us up for a claim? She said she wanted a last look before the house was sold."

"Margaret Koch was an only child. She had no children and no legal heirs."

"Are you sure? Then, why the lie?" Hollis sank back into her desk chair.

"Hollis, there are no relatives. She was probably protecting her privacy. Because of the size of the estate, I hired a private detective, Brad Pierson, to make sure there were no possible claims. After two months, he found nothing. I didn't think you needed to see the full client file, but the report is included. You may want to read it."

"Then who …. Can I see the file now?"

"Sure."

She followed him to his office.

He reached behind to his credenza and pulled a thick folder from the top of a stack. "Read it. If you discover that this young woman may be an heir, let me know immediately. No need to play detective." George straightened. "Shouldn't you be going? What time are you meeting the movers?"

"At nine. It should take about three to four hours to load everything. Then I'll go to the auction house to meet them as they unload."

He hesitated. "You look tired. When is the funeral?"

She flinched. "I don't know … I'll … I'll have to call Cathy's mother." She sat up in the chair. "Don't worry. I'll have the research you wanted done by Friday. I won't disappoint you."

George made a sympathetic face. "It never crossed my mind."

CHAPTER FOUR

❦

T HE MOVERS WERE right on time. Hollis followed them from room to room, making sure they shrink-wrapped the dressers and put heavy cushions around the glass cabinets. She walked into the den and counted all the boxes to make sure she would be able to match the count on the other end. Moving into the library, she noticed the titles on the bound volumes. She was no expert, but it wouldn't surprise her if there were a few first editions going into the many boxes of packed books.

"Excuse me, miss but this package fell from the top of the armoire in the upstairs bedroom." A muscle bound worker handed her a cardboard box wrapped in shiny silver paper, tied with string.

It had heft, but wasn't heavy.

"Which bedroom? I thought I checked all the rooms."

"Well, this was way in the back of that walnut armoire in the room with the dark green drapes. It fell down when we were pullin' it away from the wall. It's not rattlin', so I hope nothin' broke."

Hollis gently shook the box. "No, I think it's okay. I'll keep it with me."

She didn't have time to open it now. She put it with her purse in the trunk of her car. Another two hours later, the movers finally pulled away from the curb. They were going to meet her at the auction house after stopping for lunch. Hollis went from room to room making sure everything was gone. Tomorrow the cleaners would start, and the rest would be up to someone else.

Hollis chose a small eatery across from the auction house. They had comfortable booths, and she sat in one that faced the street. Suddenly she looked up from the menu, her eyes moist. She had just remembered that she and Cathy once ate here after attending the theater down the street. Her appetite gone, she took a sip of ice tea.

She reached for the box.

The paper around the box was a discolored chevron pattern lined with pale pink roses. It must have sat on top of the armoire for years. The box itself was sealed with two-inch-wide tape that ran across all four sides. She used the table knife to slice through. The tape was old and held tight, but she finally succeeded in removing it.

Letters.

The box contained letter upon letter. Envelopes with thin spidery handwriting that flowed from once navy blue ink, now faded; others were typed and equally discolored. The stationery was varied and seemed fragile, the creases a brownish yellow. There was the faintest of smells, but Hollis couldn't put her finger on the fragrance and the memory that almost came to mind.

She counted twelve letters in all.

She picked up the first one and noticed the date: 1938. She noted that the letter on the bottom had the latest date: 1957.

Out of the corner of her eye she saw the moving van pull up to the building. She carefully placed the letters back in the order she'd found them, re-taped the box, and put the money down for her bill.

This would take more looking into. She had a quick thought

about how Cathy would have loved to go through them with her. Returning the box to her car, she darted across the street. In a matter of minutes, the movers had opened the wide rear entrance doors and were unloading the goods.

A tall, slightly balding man who must be the manager said, "We have your inventory. If there are any discrepancies, we will let you know." He had an affected, almost-British accent that instantly put Hollis off. She bit her tongue and decided to choose her battles.

"Fine. I'll just stay here until they've finished."

Mr. Haughty patted his charcoal gray suit pant legs as if Hollis carried a cloud of dust with her. He was clearly ruffled that his presence alone wasn't assurance enough for her. "It really isn't necessary, but if you insist."

"I do." Hollis smiled.

HOLLIS SHOVED A file into a Fedex envelope and dumped it into the pickup box. Unless she picked up her pace, she was going to be late for her dinner with Mark and Rena. It had taken her longer than she expected to get a case file George wanted ready for mailing, and it took the rest of the afternoon to update the Koch file.

Arriving in a flurry, she was glad to see the couple looking relaxed and happy. If they were annoyed, it was well concealed.

"I was sorry to hear about your friend," Rena said. She wore a navy blue sheath and a simple gold chain with a matching bangle bracelet. Her hazel eyes, contrasted against her light, honey-colored skin, stared solemnly at Hollis.

"Thank you, I am too." Hollis took a sip of the wine they had already poured for her. "How are things going with you, Rena? You look fantastic."

Mark interjected. "Since we got together last, she's been promoted. She's now a senior buyer for Barneys. She'll be going to Paris next year for the spring shows."

"Congratulations." Hollis smiled. "Maybe you can go

shopping with me sometime. My fashion sense is … is … well, let's just say it isn't."

Rena laughed. "I would be glad to go with you. You've got a great figure and gorgeous hair."

"Thanks for the compliments. You can expect my call in a couple of months. There's this thing with Cathy, and …." She hesitated. "I just need to get through my exam results." Hollis accepted the plate of food from the waitress and waited for the others to be served. "Before I forget, how is everything with Christopher?"

"Christopher is fine. He's in kindergarten and totally brilliant." Rena nodded. "Sometimes I think back and it's so hard to believe that only a couple of years ago I came off parole—a single parent with no love interest." She smiled at Mark. "Our book club days seem so far away."

Mark squeezed Rena's hand. "That was some book club."

Hollis smiled. Rena's fraudulent check writing conviction qualified her for the Fallen Angels Book Club, and she and Hollis quickly became friends.

Over dinner they shared leisurely conversation and caught up with each other's lives.

"This fall we're taking Chris to Disneyland, and then we're off to the Bahamas for five days." Mark raised his glass in a mock toast.

Hollis toasted back. "It will be a while before I have enough time saved to take a vacation, but I'll think of you fondly."

Rena dabbed her lips with her napkin. "This was really fun, but I've got to go home to release the babysitter. I know you and Mark have to talk strategy." She gave Mark a light kiss on the lips. "Good luck."

Hollis took a sip of wine as she watched Rena walk away. "You two seem very happy. You know, I really don't feel like strategizing. Just tell me how you want to start."

Mark arched his eyebrows. "You're going to let me lead? Aren't you feeling well? Great, I do have a couple of ideas."

Hollis tried hard to follow what Mark was saying, but she was hung up on his first sentence and surprising revelation.

"What do you mean Cathy was fired from McClouds? She told me she left, but not that she was fired. What for? When?"

"Is that all you heard? Didn't you hear what I said after that?"

Hollis winced. "Sorry."

"I was only there a month before Cathy left. So that's been what—almost a year?" Mark tried unsuccessfully to flag a waitress. He turned back to Hollis. "Evidently, they let her go because one of the partners had relied on her work to substantiate a big lawsuit complaint. It had errors and was considered amateurishly faulty."

Hollis shook her head. "I don't believe it. Cathy was an excellent researcher. She always went overboard to verify her sources."

"Well, technically they didn't fire her. She was forced out with a healthy severance check."

"Why would they pay her to leave if the quality of her work was in question?"

Mark held his hand up to flag the waitress, who finally came over to their table with the bill. Hollis could feel her frustration building.

"Mark, why would they pay her?" Hollis persisted. "I got the feeling she wasn't telling me the whole truth that night, but I didn't know I wouldn't have the chance to follow up with her. Why would Cathy go from budding attorney to the uncertain life of a freelance writer?"

"I don't know, and I don't know why they gave her the money. Cathy didn't want to talk about it. You know there are no secrets in a law firm, but this was the exception." He paused. "She'd changed since she left McClouds. She was happier in one way and more stressed in another. It was different when the suit had her on edge, but ...."

"But?" Hollis prompted.

"But I got the sense that she might have taken a fall for one of the partners." Mark looked past Hollis. "She was never happy at McClouds. I think we cramped her style. She didn't seem upset about leaving."

Hollis looked up at the night sky. "I don't think we are ever going to know the whole story."

They walked out together in the cool evening. Mark walked her to her car, and she drove on auto-pilot out of the lot.

Too many questions, too few answers.

AT HOME, HOLLIS showered and climbed into bed for some sleep. She needed to hit the floor running tomorrow, checking the leads from Cathy's file. She was leaning over to turn off her bedside lamp when her eyes were drawn to the shiny box on her dresser.

She got up and gingerly removed the first letter, dated 1937.

Dear Margaret,

Things are not good here. I know you don't want to hear this, but John fell and broke his leg. He can't run the machines. They let him go at the factory. He told me not to worry, that once he's fixed up he's going to apply for one of them WPA jobs. He said they pay better anyway. Meantime things are pretty bad. But don't you worry. The thought that keeps us all going is that you are doing fine. You are, aren't you? Write us sometime.

Your Loving Mother

Hollis stared at the letter for a long moment. Looking around her bedroom she felt a chill. She was naturally nosy, but this letter was too personal and the simple words conveyed so much more than what was on the page.

Putting the letter on the bottom of the stack, she closed the box, wrapped everything back up and took it downstairs. She set it on the chair next to the door so she wouldn't forget to take it to work in the morning.

Margaret Koch had died a millionaire. How had she made that journey from what appeared to be such humble roots?

CHAPTER FIVE

———❦———

THE NEXT MORNING Hollis arrived at work early. She wanted to do some checking on the Koch matter before searching online databases to follow up on Cathy's material. She was meeting Mark in his office a little later. By then he might have heard from *Transformation*.

Closing the door to her office, she entered a search query on her computer. Kelly Schaefer's name brought no results from the law firm's person-locator database, nor did the license plate number; evidently it had been a rental. DataCheck was considered very thorough, although to do its best it required an accurate name. The screen flashed "processing" and the bar slowly crept across the file search ruler. Nothing.

After another fifteen minutes of Googling, she wasn't any closer to finding any information on the young woman who had visited the Koch house.

She went to see George.

"The name of the visitor I discovered in the Koch house was Kelly Schaefer. There was nothing in DataCheck. Does it sound familiar?" She sat down in the high-backed chair in front of his desk.

George took a moment to think. "Not off the top, no. What does she look like?"

"Early to mid-twenties, tall, pretty, brunette, dresses stylish, looks conservative."

"Nah, I'd remember her or at least her name from meetings with Mrs. Koch."

"Well, I've got a lunch meeting regarding Cathy's case. I'll keep reading the Koch letters and try to track down Ms. Schaefer."

"What let— Oh, the letters you found in the house," he said. "How long will it take you to finish reading them?"

"I'm just getting started."

George tapped his head lightly with his pencil. "Finish reading as soon as you can. They're so old that I really don't think there will be anything in them to point to living relatives, but let's err on the side of caution."

"No problem." She turned to leave and then stopped. "I can't be sure, but I think Schaefer may have taken something from the house."

"More reason to get this matter finished and filed," George said absently and began to tap on his keyboard.

Their meeting was over.

HOLLIS WAITED IMPATIENTLY for Mark to get off the phone. She was eager to get started on Cathy's case.

He finally gave her his full attention.

"I need to get back to the office." She pointed to the clock on his desk. "I've got an assignment to get a jump on. Have you got any news? What's next?"

Mark furrowed his brow. "Ordinarily I would have you pursue the premise that the information Cathy had on Fields was bogus. I'd want to know what Fields is going to hit us with. But unless you have a better idea, I'd like to know what she thought she had on him."

She nodded. "I agree. I want to get my hands on her other

research material. It had to be in her car or at home. She said she was pulling it back together, and I know she didn't leave it all with me. Were you able to get us in to see the regional editor at *Transformation*?"

"I'm happy to tell you that after many phone transfers, multiple dropped calls from my cellphone, and a couple of verifications of my credentials, yes, I finally made contact with the editor-in-chief, Carl Devi. I have never met such a group of suspicious employees. It must come from working for a tabloid. We have an appointment for this Thursday at one o'clock. But I can tell you, it took some convincing. They just want this whole matter to go away. We're going to have to show how it's to their benefit to go forward with a defense. They're ready to settle with Fields."

Hollis shrugged. "Then we have to convince them that it's cheaper if they hire us and we give them a winner."

Mark leaned back in his chair. "That's going to take some doing, but with a little luck, I think it we can persuade Devi."

Hollis pointed to her small stack of papers. "Two days isn't a long time, but I'll be ready with something. These are the pages from Cathy's folder she said she had with her when she was burgled. Before you ask, the police have the originals; I dropped them off right before I came here. Here's a copy for you. I glanced at them, but there didn't appear to be anything that was too revealing. However, I need to take my time and examine them carefully."

Thankfully, when she had taken the papers to Cavanaugh, he was out on a call. She attached a note and left them with the officer of the day. If she never put another foot inside that building, it would be okay with her.

"I'm just double checking." Mark picked up his cup of coffee. "Are you sure you want to do this thing with *Transformation*? The research establishing the burden of proof is going to be up to you. I'll help you when I can, and obviously I'll be ready for court, but I'm getting a big industrial case and it will keep me plenty busy."

"Am I sure? Absolutely, Cathy was my friend. When I was finishing law school she put in many hours to help me with finals." Hollis shook her head. "Then, when she asks for my help, I hesitate and she's gone forever. I lost a good friend. Yes, I'm sure."

AFTER HOLLIS RETURNED from Mark's office, she was motivated to draft a quick summary of Cathy's notes. It appeared as if Cathy's approach was to focus on findings that pointed to discrepancies in the financial records of Fields' charities—questionable bookkeeping procedures and even more questionable financial results. It didn't take a leap of faith to conclude that she had stumbled on enough smoke to lead her to fire—in this case, evidence for her suspicions. Fields wouldn't be the first high-profile personality who would stop at nothing to keep his good name from being tarnished. An argument based mostly on Cathy's suspicions might not hold up in court, but it might be enough to sell their services to Devi.

Then, keeping in mind her promise to George, she put aside the *Transformation* file and turned back to the Koch letters.

The next three letters were dated 1938, about two weeks apart. She read them slowly:

Dear Margaret,

We haven't heard from you for a long while now. Betsy Thompson told us she bumped into you at the Newberry's store in August. She said you looked real nice. That made us all feel good. This fall was a hard one on us. The crops didn't come in like we needed. John has all but given up on the farm. We owe everybody and I just stopped going to church. I know people are looking at us. Anna Morris and Estelle Peavey gave us a neighbor's basket. I wanted to give it back, but the truth is we need the food.

And we need you to come back home.

Paul Hitchcock said that if you did come back you could work at his store full-time. I think he still likes you. I know what you think of Rowan, but it's a good town. Everybody looks out for everybody else. It would only be for a short while. You can go back to Chicago when things settle down. You can write at home just as well as in the big city.

Let us know when we can expect you. John will meet your train.

Love,
Mother

Hollis looked out the window at the calm waters of San Francisco Bay. Sailboats dotted the eastern side of Yerba Buena Island. She now had a lead where to start looking into Margaret's past: Rowan, Illinois. She also recognized the little tingle of curiosity in her chest; she was hooked. Who was this Margaret Koch?

She picked up the next letter.

Margaret,

I got your note. It was too short to be a letter. It is clear that you have forgotten your responsibilities. I know the big city must look good to you. But we need you here.

Now come home.

Mother

Hollis grimaced, and opened another:

Margaret,

I cannot tell you the shame I felt when I read your letter. It is clear that you have no love for your family at all. When I gave birth to you after eight hours of hard labor, I didn't think twice about what an additional burden you would be. You were my child. Then when Baby Girl was born dead I knew you would be my last child.

Your dad has always worked hard to give you and Roy the best he could. He always favored you over Roy, but now this. This will kill him for sure. I don't understand when you say you have a new life as if somehow you're not connected to your old life. Are you just going to throw us away? Everybody in Rowan thinks you are coming back home to help us. What will I tell them? What will I tell your father?

Margaret Rose, I didn't want to bring this up. I know we swore that we would never talk about that night again, but I don't know how else to get you to come home. You must come home. If you do not honor your obligations, well then I don't think I'll be able to honor my word to you.

Your loving Mother

Had the threat worked? She turned to the remaining letters. The next one was dated almost a year later.

Margaret Rose,

I remember when you were born—late. We've had our rough patches but we always came through them. Every day I see your brother and sister struggle to work the land, to keep food on our table.

We got your check. I hope you will understand if I don't say thank you.

I pray to God to give me the words that will bring you home. But after all these months I can't seem to find them. I want you to know that I'm going to stop trying.

This might be one rough patch we can't get through. But I'll always remember when you were born—late.

Mother

Hollis could hear the secretaries leaving their offices

for home, but she wanted to run Margaret Koch through DataCheck. It didn't take long. When the last page printed, the file was surprisingly thin.

Margaret Koch's background check filled only two single-spaced pages. Hollis had to think back to the shortest background record she had ever read, and Margaret had that one beat by one page.

Born in 1918 as Margaret Shalisky in Rowan, Illinois, to Nora and Edmund Shalisky, Margaret was the third of three children. She graduated from Lincoln High School when she was sixteen. Her mother worked as a laundress and her father was employed part-time in a mill before he turned to farming.

Margaret moved to Chicago but came back to Rowan to marry and then, after less than a year, returned to Chicago. There was a short paragraph noting a third marriage. Hollis rifled through the pages, making sure she hadn't missed the gap of eight years or the second husband. There was nothing.

Judging from the dates on the remaining letters, they should fill the time gap. Placing the letters back in the box, she slipped the ribbon back in place and locked the box in her bottom desk drawer.

George was in court tomorrow, so she'd brief him when he returned. Hollis straightened her desk and turned out the light. As she closed the door, her thoughts sought out the silver box in her bottom drawer.

CHAPTER SIX

—·∾·—

TRANSFORMATION MAGAZINE'S OFFICES were
nothing like Hollis thought they would be. Thickly padded
Berber carpet and heavy drapes drawn back on floor to ceiling
windows set the stage for hushed conversation. Chrome and
glass furniture gleamed. A modern, oversized wall clock with
stainless steel hands ticked loudly, making everyone feel their
time was running out.

Mark motioned her to the indigo-blue leather sofa, while he
walked over to the reception desk, where a skeletally thin man
in a suit that looked to be a size too small was speaking into a
thin coil of wire. Hollis noted the high gloss of his jet black hair
and concluded that it was an acquired color. After a minute he
gave Mark a quick once-over, as if evaluating his worthiness to
interrupt his call.

He tapped a button on the console. "Yes?"

"Mark Haddan and Hollis Morgan to meet with Carl Devi.
We have an appointment."

"So you say." He went back to his call.

Mark turned to Hollis with an expression of disbelief on his
face. He turned back and picked up the name plate. "Phillip?"

he waited for him to once more put the call on hold. "I'd appreciate it if you let Mr. Devi know we're here. I wouldn't want him to think we didn't arrive on time."

Hollis barely held back a laugh. With another punch of the button the young man turned his back to Mark. She caught her breath. Phillip's movement put him directly in line with the light from a halogen lamp; the man's skin was pulled tight over tiny broken veins. His eyelids were stretched almost to the point where Hollis could barely see his pupils. His surgically perfected lips resembled swollen slashes in pastry dough. He should definitely stay away from any light source. It was clear that despite his best efforts Phillip was a fifty-year-old man who was losing the battle to keep his youth.

"You may have a seat," he said, his back still to Mark.

"Good grief, what did you do to set him off?" Hollis asked when Mark settled in beside her.

Mark shrugged. "Hey, it wasn't me, but the magazine could use a better first impression at the front desk."

Moments later Carl Devi strode into the room.

"Mr. Haddan, Ms. Morgan," he said with an outstretched hand. "Let's go in the conference room."

The conference room was every bit as impressive as the lobby. Slick and stylish, Devi cut a fine figure as he took the seat at the head of the high gloss walnut table. He wore a crisp white shirt, silk maroon tie, and matching suspenders. Before he sat down, Hollis had noted pin-striped navy slacks cuffed over shiny black shoes. The Rolex on his right wrist was understated but not modest. Dark hair with shocks of gray at the temples made him appear both distinguished and aloof.

"Can I get you some coffee?"

"No, nothing for me, thanks." Hollis chose the chair that allowed her sit with her back to the window.

Mark declined as well and took a chair across from her.

Mark leaned in and folded his hands on the table. "Mr. Devi, we know you're busy. So I'll get to the point. Catherine Briscoe was a good friend. Actually, we once worked for the same law firm. She and Hollis attended the same law school." Mark paused. "We believe in her work and we want to redeem her name."

If Devi was surprised or even interested, he didn't let on.

Mark pushed ahead. "Hollis is a top paralegal, and while my area of law is largely corporate, I am comfortable representing either the plaintiff or the defendant in libel matters."

It was Hollis' turn to speak. "Mr. Devi, Mark and I would like to prepare your answer to the Fields' complaint."

Devi turned to face Hollis.

"We would take on the research and work with your attorneys, of course," she added. Hollis still couldn't read Devi's blank expression. She decided to wait for him to comment.

Mark, on the other hand, pushed his chair back and stood, determined to soldier on. "Look, I know that the lawsuit is basically moot. *Transformation* can file for dismissal since the offender is dead and the article was stopped, but—"

"But, we will work for free to clear her name," Hollis broke in. "We need standing in the court. *Transformation* will just need to cover our court and filing fees so our firms won't be out of pocket."

Hollis regarded Devi's silent demeanor with some impatience. He just sat there with what looked like a hint of a smile.

Devi looked from one to the other. Finally, he slapped the table with his palm

"We want your story."

"What story?" Hollis was taken aback.

Devi got up and started to walk around the room, stopping next to Mark. "Yes, this could be a great story. A poor struggling lawyer turned writer is obsessed with a fantasy about a popular celebrity. She risks it all but fails. Her loyal friends take up the banner to salvage her reputation. Will they meet with the same disastrous end?" Devi took out his phone and started madly texting.

Mark spoke. "Mr. Devi, I don't think it would be appropriate to …."

Devi squinted. He obviously didn't like anyone to interrupt his momentum. "You want our help, or not? Leave your card. I'll get back to you."

Hollis motioned to Mark to head for the door.

"We'll look forward to hearing from you. Not about our story … about the complaint," Hollis said, moving to the door.

Carl Devi waved goodbye and pushed the door shut.

As they waited for an elevator, Hollis shook her head. "What was that all about?"

"That was strange," Mark said.

"I just can't imagine Cathy working for that guy. He's so …." The elevator arrived.

Mark nodded. "If we get the job, we'll make sure we have at least one serving of caffeine before any meetings."

HOLLIS CLOSED HER office door and forwarded calls to the reception desk. She had passed on lunch with Mark and come straight back to the firm from their meeting with Devi. It had been five days since Cathy's death, and the pit in the bottom of her stomach was still there. Time was slipping by, but she had promised Mark not to start digging until they heard back from Devi.

Her phone rang and she snatched up the receiver.

Mark's excitement was tangible. "Hollis, I just heard from *Transformation*. They'll allow us to work their claim."

"That's great, but will they cover our fees?"

"If we win they'll give us a twenty-five percent bonus and reimburse our 'reasonable' expenses." He paused. "They also want us to brief them on a regular basis. If we lose, there is nothing. They're going to assign us a senior attorney, who'll also report our progress and get back to Devi about our findings."

"Okay, I get it. It's not a problem to keep them in the loop. We aren't doing it for the money anyway."

"No, we're doing it for Cathy," Mark said somberly. "I'm done for the day. I'll file the papers for a change of counsel with the court in the morning. Then I'll see if we can get a sit-down with Fields' attorney next week for a continuance."

"Now, that will be a hard sell." Hollis reached for the small file of papers. "I'll start researching the corporation papers for Fields of Giving. It's a long shot, but we need to check off the boxes." She flipped through the sheets of paper in Cathy's file. "I would have expected more factual articles, rather than just these puff pieces for someone as prominent as Fields."

"Yeah, I know." Mark sounded worried. "I took a look at the contents earlier."

"Still, not a problem. If anything is there, I'll find it. But first I've got to get a filing out for George." Hollis moved the box of letters to the side of her desk. "You focus on *Transformation*. I'll work the background research. Let's talk tomorrow."

"Sounds good, our number one goal is to get that continuance."

"And I'm going to contact Cathy's mother to find out about the funeral and see if she needs anything. I'm hoping she'll talk with me."

"Does she know you?"

"I met her once with Cathy. It was only for a moment, though. I don't think they were that close."

"I know it might be a difficult conversation, but it can't hurt to see if she knows something."

Hollis frowned. "I'm not very good at these kinds of things."

"Don't sound so worried," Mark said.

Her shoulders slumped. "I'll call you later."

IT TOOK MOST of the afternoon to finish a case filing for George. Over the rest of the day she prepped two more files for his review; then she was ready to leave. The street lights had started to come on.

***

AT HOME SHE placed her keys on the entry table in the mouth of one of the porcelain frogs she inherited from her grandmother. She would shower, fix a small salad, and call it an evening.

Walking heavily up the stairs, she stopped midway.

A rush of tears blurred her sight, and she had to sit on the staircase and give in to her quaking shoulders and the unceasing ache in her chest.

She said goodbye to her friend.

CHAPTER SEVEN

❧

HOLLIS DREADED MAKING the phone call. She did as much desk work as she could until she couldn't put it off any longer.

"Mrs. Briscoe, my name is Hollis Morgan. I was a friend of Cathy's."

Even over the wires, she heard the intake of a sob.

"I'm sorry. I don't know if you remember me? I wanted to know about the funeral and …." Hollis' words were rushed. "I'm sorry. Maybe this is a bad time."

"No, no, it's just that people keep calling. I didn't know Cathy knew so many people. The funeral is on Wednesday, but I think I remember her bringing you by to meet me. You have an unusual name." She broke into sobs again.

"Mrs. Briscoe, Cathy and I were good friends. Is there something I can do to help?"

"No, my sister is flying in today." The thought seemed to calm her. "But thank you for your offer. I'm sorry I can't seem to stop crying. It's just that Cathy was a good daughter. I don't know why anyone would kill her." She blew her nose. "I'm sorry, why did you say you called?"

"Mrs. Briscoe, did you say Cathy was killed?" She shifted the phone to her other ear.

There was another sob. "Yes, the police said Cathy was drugged with sleeping pills before she was placed in the car."

Hollis frowned. "But she had a prescription …. She told me she hadn't been able to sleep."

"I know, but last night the detective called and said the police report found another drug in addition to her own sleeping pills, another strong … stronger …." Her voice broke. "I'm sorry, I really can't."

So Cathy *was* murdered.

"No, Mrs. Briscoe, I need to apologize. I didn't mean to bother you. Like I said, I wanted to offer my help and to come by and talk with you. Cathy came to see me the night before … the night before she died."

There was another sob.

"This isn't a good time. Mrs. Briscoe," Hollis said. "I'll contact you next week and—"

"No, the funeral is Wednesday." She took an audible breath. "I want to talk about her. No one wants to talk to me about her. Her father is too busy to come to his own daughter's funeral. Says he can't bear to see her in the … in the ground …. Her brother is in Afghanistan. I can't reach him. Please, please I would like you to come and talk to me, a friend of Cathy's. Come around ten o'clock tomorrow."

EVELYN BRISCOE LIVED in what appeared to be a quiet middle-class neighborhood in Castro Valley. Leafy maple trees lined a cul-de-sac with large front yards and two-car garages. Neighborhood Watch warnings were posted on both sides of the street. Hollis didn't mind giving up her Saturday. She walked up the curving pathway to the smallest house in the circle.

After awkward greetings, Hollis chose to sit on a tufted loveseat in the cluttered living room. Evelyn Briscoe had aged well. From a distance, she could almost be mistaken for Cathy. She wore her dark gray hair pulled back into a ponytail. Brown tortoise-shell glasses made her look stylish, sophisticated, and somehow down-to-earth.

"Of course, now I remember you, Hollis, from when you visited. Cathy would also mention your name from time to time." Evelyn sat on the large L-shaped sofa. "This whole awful …. Cathy was always so headstrong and so sure she was right." Evelyn Briscoe picked up a picture of her daughter from the top of a stack of other framed pictures that covered two sofa cushions. "So … so righteous, just like her father."

Unsure how to respond, Hollis gazed around the room filled with early American memorabilia.

"I'm trying to follow up with Cathy's defense against Dorian Fields. I'm working with an attorney who was also a friend of hers. We're trying to take up her defense in a libel suit. But I need to ask you a few questions, and I want to know if you would mind letting me have the keys to Cathy's condo."

"Libel suit. That's Dorian Fields isn't it? He was the one suing her. She said he was a crook. She told me she had proof."

Hollis tried not to look too hopeful. "Did she say what it was?"

"No, she never would say."

"Can you think of anything Cathy said about Dorian Fields that might help us?"

Evelyn Briscoe's lip started to quiver and her already red-rimmed eyes started to glisten. "You think he killed her, or maybe had her killed?"

"We don't know. But we feel we owe it to Cathy to discover the truth."

"The police were here yesterday, asking the same questions." Evelyn stared out the window. "She was so smart. She made me feel stupid. But I was so proud of her. She let me keep all her awards, all her certificates. She said she didn't need them. She said …."

Her voice drifted off into the silence of remembrance.

Hollis shifted in her seat, and the movement caught Evelyn Briscoe's attention. She started speaking again.

"One time we were having coffee in the sunroom. It was hard to get her to slow down. She was in a good mood and kept checking her cell for a call." She took a long breath. "When the call came through she was happier than I had seen her in a long time. She gave me a big kiss on the forehead."

"When was this?" Hollis asked.

"About a month ago."

"Did she say anything about the call? Was it about Fields?"

"She didn't say, but I got the feeling it was. Work was the only thing that made her happy."

Hollis reflected that most people could say the same thing about her.

Evelyn started to tear up again. Hollis reached over and held her hand.

She smiled weakly. "You came for her keys. I'll get them." She paused. "Would you mind … I mean while you're there … just her personal things … I can't bear to visit where …."

"Of course," Hollis said, patting her hand. "I'll bring them to you. Would you like me to pack up both her bedroom and bath?"

"No, no my sister will do that and my brother will pick up any boxes," she murmured. "I'll get movers to pack everything else. Not now … later."

"All right, it's not a problem."

It was time to go.

"Mrs. Briscoe—"

"Wait, I do remember one thing she said after that call. It was odd. I didn't understand it at the time and Cathy didn't explain. I didn't want to break her mood, so I didn't ask." Evelyn got up went into the kitchen and returned with the keys.

Hollis pocketed the keys but remained silent, not wanting to distract her from recalling the memory.

"She said: 'To be a great fisherman, it is important to know how to bait and switch.'"

WHEN SHE GOT back to the office, Hollis checked that she had no messages. The visit with Evelyn Briscoe had depleted her. Shutting her office door, she settled in her desk chair and closed her eyes. Her own relationship with her mother came with a skull and crossbones label. She had no doubt that her mother would, in similar circumstances, blame her.

Opening her eyes and taking a deep breath, Hollis decided that she needed a distraction. She pulled the Koch box out and read the next letter. It was written in 1939.

Maggie,
Mama died two weeks ago. She never told us where you were. We found your letters when we was packing her things. She was sick for a long time. Jenny and I knowd there was a reason why you up and left. But you can come home now. It's just us. And we miss you. We are sorry.
Love from the heart,
Roy

According to DataCheck, Margaret never moved back to Rowan. Roy was Margaret's older brother by four years. There was a Pierson file cross reference to a Roy Shalisky. From the date of the letter, he died in a flu epidemic just over a year later.

She opened the next letter.

Margaret,
I know you told me not to contact you, but I needed to say some things. You are so young and pretty I don't blame you for leaving an old man like me. I don't know art and I don't know music and things, but I do know I love you. It doesn't matter about the money. I told you it was yours. I hope it makes you happy.

People say I'm an old fool, and that you used me. But you and I know different.

Have a good life.

Yours,
Granger

Hollis flipped the envelope over. A Chicago postmark, dated 1940. Margaret would have been twenty-two years old when she received it. She was suddenly very curious to find a picture of Margaret Shalisky Hitchcock Ferris Koch. She must have been attractive. Hollis reached for the private detective's file. There were two color photos, but Margaret looked to be in her sixties or seventies. Even so, she was a striking and handsome woman. It wasn't too hard to imagine a young and vibrant Margaret.

There were several black and white photos clipped to the back of the folder. Hollis quickly scanned through what appeared to be family members gathering for different holiday occasions. One photo caught her eye: a young girl of about three or four stood next to a seated, middle-aged Margaret. The face bore a faint resemblance to the young woman, Kelly Schaefer.

But then that was impossible, or so the investigative report would imply.

Hollis wanted it to be the end of the day, but only a couple of hours had passed. She was feeling uncommonly emotional and she didn't know what to do.

She read the letter, dated 1941:

Dear Margaret

I am glad to hear you are doing well. You said those business courses you took would give you a head start, and I guess you were right—an executive secretary, my goodness.

Well Tim and I are getting along better, especially when he found out we were going to have another baby.

I hope it's a girl so she can share the room with Laura. If it's a boy, well the other boys will just have to move over. I hope this is our last one. Doc Riddle said that we need to space them better. Actually he said I shouldn't have any more, but I didn't tell Tim. He likes me pregnant. But I have to tell you I am so tired these days. I can't seem to catch my breath.

I was in the post office yesterday, and everybody was talking about how all the jobs are gone with the men leaving their farms to go to war. Tim is deaf in one ear so the Army wouldn't take him. The depression is lasting longer than we all thought, for us anyway. Well I've got to go fix supper.

To answer your question, Paul Hitchcock is still single. He moved next door to the Potts farm. After he got that big inheritance from his grandma, every girl in Rowan has been throwing themselves at him. But I think he's waiting for you to come back. Any time anybody talks bad about you he gets upset and tries to defend you. I try to as well, but I was your best friend. Paul is sweet. Are you coming back, Margaret? Write me.

Your loving friend,
Miranda

She replaced the letter just as George knocked and entered her office, coffee in hand. "How's the case going?"

She frowned. "Not as well as we would like. *Transformation's* attorney is essentially using us as his clerical pool. Mark and I haven't found anything that even looks like a viable direction." Hollis sighed. "*Transformation* management already requested the documents we need for discovery, but the gut of it, the proof, is still elusive. Our strategy is to interview non-profits who benefited, but you can imagine they'll be pretty closed mouthed—obviously. In fact we're meeting this afternoon with Fields' team to ask for a continuance."

George wrinkled his brow and held up his hand. "I meant with the Koch estate."

"What? Oh, sorry." Hollis blinked a few times. "It's going a little slow, but I'm determined to knock out these letters in the next couple of days."

Hollis could tell that George had something he wanted to tell her. He looked at her over his glasses. "I got your request for a subpoena. What do you think you'll find in Margaret Koch's health records?"

Hollis was prepared to respond, but she didn't think it was a good idea to tell him that her request was pretty much based on curiosity and unanswered questions. "I'm going through the letters, and there's a question I had about the health of her second husband."

"Let's hold off on the subpoena. I think it's going too far." George removed his glasses. "*Second* husband? How many did she have?"

"Well, at least two, but I'm not quite halfway through." She picked up her notebook. "One other thing, George, did you ever meet Margaret Koch?"

"Yes. I did her last codicil about three years ago when she was eighty-five. She was articulate and gracious. I remember she had a somewhat coarse sense of humor. Why?"

Hollis avoided his gaze. George knew her too well and would wonder why she was dragging her feet with a subpoena request. "No reason, really. Reading her letters I'm starting to have a picture of her in my mind."

"She didn't—"

His cellphone rang. He motioned he was leaving, and pointing to the Koch file, mouthed, "Hurry up."

Hollis smiled. "Don't worry, George. Like I said, I'll have the letters finished before the end of the week."

He waved a goodbye.

She gave him a smile, took the letters out of the drawer, and opened a business envelope dated November 1942.

Dear Mrs. Hitchcock,

    Apex Insurance has approved the claim you submitted to receive the benefits of your late husband, Paul Hitchcock. However, due to the brevity of your marital union and since Mr. Hitchcock did not name a beneficiary, we are withholding disbursements in the amount of $5,000 for 120 days, until we are assured that no other claimants will come forward.

    Please do not hesitate to contact us if you have any questions.

Sincerely,
Phillip Barnsdale
Adjuster

Well, ol' Margaret didn't waste any time snatching him up.

Hollis went into her desk drawer and pulled out the detective's report. This was Margaret's first marriage. She glanced at the marriage date—only three months married to Paul. She looked through the file, flipping to Paul's age at death. He had gotten the flu and then pneumonia. He had been only thirty-two to Margaret's twenty-four.

Next, she picked up rose-colored note paper tucked inside another letter. The note appeared to have been scribbled hurriedly:

Dear Maggie,

    I was able to get you an invitation to Jenny's party, but please do not bring Charles the BORE with you. Don't you know any young people our age?

Rosemary

The note wasn't dated, but the letter was written in 1943:

Dear Margaret,

I can't tell you how much I treasure seeing you these past months. You were very kind to me and it helped me to forget for a while the loss of my wife. I never thought I could be happy again, but you have proved me wrong. Last night awakened in me a man I thought I had sent away.

My enlistment papers have finally been approved. Perhaps the Army has lowered its standards. I don't know if I will come back from this damnable war, and I won't ask you to wait for me. But if I should return, I promise you that I will do everything I can to make you mine.

I've told Eric to keep an eye out for you. You're not as independent as you think you are. Go to my brother if you need any help. He is upset that his asthma is keeping him out of serving, but I'm glad he's out of harm's way, and even gladder he will be there for you.

I love you, my darling. There, I've said it. I only wish I could see your face.

Embracing You,
Charles

The next letter was dated, 1944, almost a year later.

Dear Margaret,

At first I was distraught that you had returned to Chicago. But then I could understand that Rowan was not a good place for an ambitious young woman. I'm sorry Eric wasn't much help. He hasn't written me, or at least I haven't received any letters. Yours only caught up to me two days ago. I can't tell you how much they cheered me.

I'm tired. We are told the war in Europe is going in our favor, but all we see here is rain, mud and mosquitoes. The one thing that keeps us all going is the fact that the other side is worse off than we are.

I can't write more. I want to make the mail pouch.
Please wait for me. I need to know you that you are in my
future.

Always,
Charles

The paper was grayish-yellow, and Hollis thought she could
smell the dankness of the battlefield. Checking the time on
her computer, she remembered that she was meeting Mark at
Fields of Giving headquarters and didn't want to be late. She
opened the next envelope and pulled out the letter.

From its folds, a yellowed newspaper clipping with the year
1945 scrawled in the corner fell in her lap.

Maggie,

I am really sorry to tell you this, but Charles returned
from the war yesterday. They brought him in an ambulance,
and you know for Rowan that was a very big deal. Charles
is a hero, Maggie. He was awarded the Medal of Honor
for saving his squad from a grenade. Unfortunately, it cost
him his leg and an arm. He wanted me to tell you. He says
that he would understand if you didn't want to be saddled
with an "old cripple." Those are his words, not mine.

He wants to see you. His spirits are pretty low. When
will you be coming to Rowan?

Your friend,
Rosemary

She picked up the clipping:

Eric Ferris / Margaret Hitchcock
The Chicago Review
Chicago, Illinois
Saturday, June 5, 1945
Chicago—Mrs. Margaret Hitchcock, of Rowan, Illinois

was wedded here to Mr. Eric Ferris of Rowan, Illinois,
on Saturday afternoon by Reverend C.H. McCoy at the
Second Presbyterian Church.

Hollis leaned back in her chair. Margaret had no qualms
about seeing two brothers and marrying the one without
disabilities. She didn't want to judge, but she wasn't sure she
liked Margaret. The more baffling question was, why did
Margaret keep the letters? The condition of the envelopes
seemed to indicate she had re-read them at least once.

One thing Hollis was sure of: she was done reading letters
today.

CHAPTER EIGHT

H OLLIS AND MARK entered through double French doors into a broad foyer that fronted the lobby of the Fields of Giving, Inc. headquarters. Unlike the Transformation offices, which were cool and sterile, Dorian Fields' place of business had a warm atmosphere, with rosewood floors, piped-in new age music, overstuffed velveteen sofas and a large fire pit table with low but steady flames.

She put a hand on Mark's arm to indicate that this time she would take a turn with the receptionist. They both walked up to the young woman greeting them with a broad smile.

Hollis smiled back. "We're here to see Wade Bartlett."

The receptionist was just about to tap the phone button before a booming voice greeted them.

"Wade Bartlett, Ms. Morgan, and Mr. Haddan, welcome." He entered the room with his hand extended and a cheerful grin. "Come on back to my office; we can talk there." He led them down another wide walkway.

He was a tall blond man with blue eyes and horn rimmed glasses that gave him a kindly professorial look. His navy blue cable shawl sweater draped casually on his lean frame and complemented his conservative striped tie.

Mark settled into one of the two thick green upholstered chairs. "Thank you for taking the time to see us." He pulled a sheaf of papers from his briefcase. "As you know, we are the new co-counsel for *Transformation*. We're doing a little catching up, but I .... We wanted to meet face to face to discuss your position."

Bartlett started to laugh and then caught himself. "Our *position*? Mr. Haddan, your client—and I was sorry to hear about her suicide—was irresponsible and reckless. She attacked a man who has given his life for those who have no hope. She—"

"Mr. Bartlett, it seems you haven't heard," Hollis broke in. "Catherine Briscoe was murdered."

Bartlett tilted his head. "Murdered?"

Mark nodded. "The police aren't broadcasting it, but you have to admit Mr. Fields makes an awfully good suspect."

Bartlett bristled. "Except for one thing, Haddan, she was lying. Where's her proof?"

Mark and Hollis both stiffened.

"Whoever killed her took her proof. But we're piecing it back together. They didn't get everything." Hollis moved to the edge of her seat. "Cathy was a good attorney and a professional writer. If she said Fields was ... dirty, then his days as a public icon are numbered."

It was Bartlett's turn to stiffen. "Well then, counselors, sounds like you're on your way. What do you want from me?"

Hollis and Mark looked at each other. Mark took out another sheet of paper and passed it over the desk to Bartlett. "We need a six-month continuance. If you agree, the court won't hesitate to grant one."

This time his laugh continued for some time. Hollis gazed up at the ceiling until it came to an end.

Bartlett picked up the paper and tossed it aside without looking at it. "Why would I do this? You're accusing my client of being a fraud and possibly a murderer. Each day that goes by, this thing is hanging over my client's head, tarnishing his reputation. Sorry, we can't help you."

Hollis made an effort to control the tone of her voice. "Mr. Bartlett, we don't blame you. But like we said, reasonable people like the twelve who sit on a jury might think your client is a prime suspect." She opened her notebook to a page of notes. "You agree to this continuance and it says he has nothing to hide. It says Fields wants the truth to come out, that Fields of Giving is a respected organization, it says—"

"It says I'm crazy," Dorian Fields entered the room with an energy that made everyone sit up.

Bartlett immediately rose and pointed to his chair for Fields to sit. Fields ignored him and pulled up a chair next to Hollis.

Bartlett spoke up, "Hollis Morgan, sir, and this is Mark Haddan, and they're attorneys."

"I'm not an attorney, I'm a paralegal." Hollis reached out her hand to shake. She was determined not to look as nervous as she was feeling.

Fields was not a tall man but he had what many tall men lacked: bearing. Dressed in a two-piece deep green suit, he filled the room with his presence. Hollis gave Mark a pointed look as they shook hands all around. With his full head of wavy white hair, Fields looked to be in his fifties, although Hollis' research had revealed he was actually seventy-two. Known to be a fanatical swimmer, he had a toned physique and smooth, tanned skin.

"Well, it took some nerve to face me on my own turf," Fields said.

Wade Bartlett pushed the request for the continuance across desk to Fields.

Fields glanced down. "Despite what your friend said about me, Ms. Morgan and Mr. Haddan, I am not a crook and I have nothing to prove. But *you* do."

Mark gave Hollis an imperceptible wave-off. He wanted

to respond. "Mr. Fields, we have just been added to team defending Ms. Briscoe. We need more time to prepare. We are hoping you will be fair."

Hollis could tell from Fields' amused look he was not moved. She turned to face him directly. "We won't lie to you; we think you could be guilty of fraud, fraud on the group of people who can least afford it. But we both know that without Catherine Briscoe's case paperwork we are at a grave disadvantage. If you are innocent, then we will prove that too."

Fields picked up the paper and scribbled his signature. "You've got sixty days, Ms. Morgan."

HOLLIS ALMOST RANG the doorbell to the condo before she remembered Cathy would not be answering. She rustled through her purse and pulled out the keys Evelyn Briscoe had given her. Inside, the sole illumination in the living room was a sliver of sunlight peeking through the drawn drapes. The room smelled of the stuffiness that comes from not having an occupant to circulate the air. Hollis pulled back the curtains, and the harsh glare made the room look even more desolate. She dumped three empty boxes in the living room and took two into the back bedroom, which Cathy had converted into an office.

The office, disheveled from the police search, had fingerprint powder everywhere. Evelyn called her that morning to say the police had released the apartment and removed the crime scene tape. The thought of what Cathy must have gone through made Hollis shiver. And if she let her mind run with that thought, she could imagine there was a chance Cathy knew her killer.

Everything had been taken apart and haphazardly put back. Pushing aside her feelings of dismay, Hollis went through the room and quickly filled two boxes.

Next she went into the master bedroom, where Cathy's interior design talents were evident. Hollis remembered when, after months of searching, she had purchased her dream comforter and matching bed skirt. The soft, dove gray and ivory contemporary pattern was complemented by a floral mauve area rug. She went through the dresser and began placing Cathy's clothes and jewelry in packing boxes. She was surprised that it didn't take long at all.

A life could be so easily packaged.

Hollis sat down on the edge of the bed. The police must have been satisfied with rifling through the dresser and closet; the rest of the room was only in mild disarray.

"Talk to me, Cathy," Hollis said out loud. She looked around slowly. A picture on the dresser caught her eye. It was all of them: Cathy, Hollis, and Marla at the Marriott Courtyard in San Francisco, drinks in hand, smiles all around. The three of them used to be close. They weren't constant companions but they were life allies. Hollis felt a rare soft spot for the friendship. That day they swore to always be there for each other, no matter what man was in their lives. That had gotten a laugh, since none of them were in any significant relationships. Two years later, Marla was still single, practicing law in Greece, and Cathy—Cathy was dead.

Shaking off her dark thoughts, Hollis jumped up and returned to the office. They'd all shared their house keys with each other, except Hollis, who didn't want her friends' keys because she didn't want to give out her own key. She reasoned that her prison time could complicate things. She didn't fool them.

Hollis stopped in the middle of the room and looked at the windows.

Tossing two floor pillows onto the bed, she crossed over to the larger of two window sills. When Cathy had her condo renovated she got her contractor to install a safe in her window sill. She once said that her contractor was the only real significant man her life. They had all laughed. She acknowledged that it wasn't a real safe, but it would hold valuables, personal documents, and her working papers. Hollis had never seen it open, but Cathy said it turned out to be a great idea. It was only an inch or two wider than the other sill. There was no way the police, let alone robbers, would know of its existence.

Too bad Cathy didn't tell her friends how to get inside.

Hollis banged on the end of the sill with her fist, but nothing happened. She went to the other window and did the same. Nothing. Cathy loved puzzles. She and Hollis would send each other encrypted emails—they both loved a Sudoku challenge.

She would have hid her Fields of Giving working papers.

Hollis tapped all along the edge of the window sill with increasing force. She slid her fingers underneath the wooden ledge, seeking a button or lever.

"Think, Hollis, think."

Then turning back to the sill she put her hands on either side and lifted. The resulting pop almost threw her off balance.

"Yes!"

Digging into the narrow, plastic-lined space, she pulled out a passport, a large white envelope, and finally a manila folder containing another oversized envelope marked "pending."

Hollis placed everything in piles on the bed. The white envelope contained Cathy's will, her birth certificate, and an insurance policy. Inside also was a tiny jewelry case with diamond stud earrings. She would take these items to Evelyn Briscoe.

In the folder was a bundle of notes and newspaper clippings about what appeared to be the topic of Cathy's next story. Her heart sank.

No references to Fields. She would go through the clippings and notes later. Hollis gave a disappointed sigh. She was stacking the papers together when a picture fell out of an envelope. It was Cathy with a middle aged woman who looked as if she had seen some hard times. Hollis flipped it over. The words surprised her: "You saw me for who I could be and not what others said I was. You said it wasn't a big deal for you because you could, but it was everything to me. You gave us our lives back. Thank you, Gloria."

Cathy never mentioned helping anyone. Hollis dabbed at the moisture in her eyes and placed the picture in the stack to go to Evelyn. Mrs. Briscoe might find some solace in seeing this side of her daughter.

She tucked the papers into her briefcase and replaced the sill. For the next couple of hours she packed and bagged what she could and labeled contents on a sheet of paper that she taped to the sides of the boxes.

With a last glance around the room, she slowly closed the door behind her.

DESPITE THE COLD foggy morning, Cathy's funeral was well attended. Hollis hated this societal tradition, but she felt obligated to go. She gave an acknowledging nod to Mark, who appeared to have come with others from his office. Thankfully, it was over quickly. Cathy's friends gathered only for a moment after the service, knowing that Cathy would be the first one to tell them to get the hell out of there.

Even so, Hollis was running late for her meeting with the detective, Brad Pierson, who had conducted the Triple D search for the Koch heirs. Through the Hayward hills, she drove just under the speed limit to negotiate the winding curves of Mission Boulevard. Hayward was primarily a middle to working class city in the East Bay Area with some of the best views of San Francisco. Another time she would have pulled over to take in the scenic vista, but today she was eager to talk to the man who might give her more background for the letters. There was something about the letters that drew her in like a heat seeking laser.

When she told George of her visit to the Pierson offices, he cautioned her, "Let Brad know these letters you found might be troublesome. Do what you need to clear them as soon as possible. I don't have a problem with you working with Mark to fight the Fields suit, but I don't want you sidetracked from the Koch case. Give it your full attention."

"I understand," she reassured him. "I can do both."

Pulling up in front of the non-descript building housing Pierson Investigations, she noticed the slightly worn sign sitting in the slightly dirty window. It fit in well with the rest of the buildings on the street, half of which had "For Lease" signs on the façades.

She went inside.

The lobby décor was just a little less shabby than the exterior. The long walnut reception desk in front of a wall-sized mirror was unoccupied. In the middle of the marble floor stood a sign on a stand with a large arrow pointing to a directory on the opposite wall.

Pierson Investigations was on the second floor. Hollis took the stairs.

The glass entry door bore a "Come In" sign, but there was no one at this reception desk either.

"Hello," she called out.

Nothing.

"Hello." A little louder. This time she heard a distant door closing, followed by footsteps, and the entrance of a tall brown-haired man. He was pulling on a jacket and slipping earphone buds out of his ear.

"Sorry," he said. "My receptionist is out for a few days, and I lost track of time. You must be Hollis Morgan."

"Yes," she said extending her hand. "What are you listening to?"

He laughed. "The game. Let's go back to my office." As they walked, he added, "On the phone you said Dodson, Dodson and Doyle had some questions about the Margaret Koch report?" He gave her a wide, handsome smile.

"Sort of. I'm a paralegal and I was sent to inventory Mrs. Koch's belongings. In one of the rooms tucked on top of an armoire was a box of letters."

"Tell me more about these letters. Do you have them with you?" Brad Pierson's gaze sought and held hers. He reminded Hollis of a history professor whose class she had taken in college, except her history professor didn't flirt with her.

Hollis coughed to break his gaze.

"No, I didn't think to bring them. Can you answer a couple of questions for me? I've read your report."

He shrugged a go-ahead.

"I read that Margaret's brother and his child are dead, as well as her younger sister. But her mother was the youngest of five children. Didn't Margaret have any cousins?"

Brad reached for a bottle of water. "Rowan, Illinois, was hit hard by the 1918 flu epidemic." He swallowed deeply. "Margaret Koch's father died that same year. Her mother died shortly after. There was a fire in the courthouse in 1930 and a lot of records were destroyed. But if you check the public record there were no relations still alive after World War II. "

"You verified all their deaths?"

Brad re-engaged his stare and didn't answer.

"Of course you did." Hollis looked down at her notes and accepted the bottle of water he handed her. "There's a gap of twelve years," she glanced down at the report, "from the age of sixteen until she shows up in Chicago. Wasn't there any information about where she lived during that time?"

"Hollis, can I call you Hollis?"

She nodded with impatience.

"Well, I wasn't doing a genealogical search. I verified whether she had heirs. I assure you, during that period of time, she didn't have any children. That was all I was looking to confirm."

"Did you run across anyone named Granger in your investigation?"

Brad's brow wrinkled. He tapped the keys on his keyboard. "No, no Granger. Who is he? Was he mentioned in the letters?"

"Yes, they may have had a relationship. I haven't read all the letters yet, so I can't be sure."

He looked at her, waiting for her to continue.

Hollis took another swallow of water.

"Anyway, I just wondered if he was important to her. That's all I wanted to know." She opened to a page in her notebook. "Oh, wait, one more thing, the pictures. Were these all you could find?"

Brad folded his hands on the desk. "Yup, and I was only able to get these from Mrs. Koch's housekeeper. Mrs. Koch's only visitors were friends ... like I said, she had no family."

Hollis laughed. "I get it ... no family." She stood. "But ... well, I did run into a Kelly Schaefer. She said her family was close to Margaret's."

He straightened. "Schaefer? Sounds familiar."

Hollis took out the picture and pointed to the young Kelly. "That's her. She hasn't changed much."

He pulled the picture over to him. "Yeah, I remember looking her up. She was raised by her grandfather. I checked out all the people in that photo and no one was related to Koch. Did Schaefer say she was?"

"No, she's been to the house, but her connection doesn't seem to be a blood one."

"Good, then my reputation for knowing what I'm talking about stays intact."

She stood. "Thank you, Brad, for meeting with me. I hope I can call you if I have any other questions?"

"Absolutely." Brad shifted from one foot to the other. Then, noticing she was ready to leave, he walked with her toward the door. "Hmm, Hollis, if you're available, I'd like to ask you out sometime."

She was glad she wasn't facing him. As she always did when uncomfortable, she coughed. She turned to confront his teasing smile. Not that she didn't have offers .... She did go out occasionally, but not with anyone she could take seriously. For some reason, she felt as if Brad might be an exception.

"I'm not sure that would be a good idea."

"Why not?"

Hollis smiled, "Suppose I find a missing heir?"

"You won't."

She smiled and reached into her purse. "Here's my card."

He gave her a broad grin, revealing a dimple in his right cheek. "Great, I'll be calling you … soon."

Hollis sat in her parked car and checked the time on her iPhone. She checked herself in the rearview mirror to see if she still had a silly grin on her face.

She had an offer of a date.

While it might be unlikely she would find a missing heir, she had to admit she'd take perverse pleasure in knocking Brad's self-assuredness down a notch.

She debated picking up a quick sandwich before going to her next appointment with the Heaven's Praises administrator—number one on *Transformation's* list of non-profits. Hollis and Mark weren't sure if that placement meant good or bad.

Hollis put off eating until she could take her time. She would finish the interview first.

PULLING IN FRONT of Heaven's Praises, Hollis was impressed with the modern store front, and even more, the off-street parking. As she entered the small foyer, a tall, middle-aged woman approached her with her hands shoved deep in the pockets of her smock.

"Call me Joy," she said. "We don't use last names here. Come on back to the community room."

Joy appeared to be in her forties, with a warm and motherly smile. She walked with a slight limp, and her fingers were gnarled from what appeared to be arthritis. She wore her once blond hair, now gray, close to her scalp in a pixie. Hollis kept pace behind her.

"What kind of services do you provide here?" Hollis asked.

"A couple of days a week we get someone in from the Salvation Army to talk about twelve-step programs." Joy looked back over her shoulder. "Then a nurse practitioner comes once or twice a month. But mostly we give street people a safe place to sleep and get their mail."

"Is it possible to tour your facility?"

"No, we respect people's privacy."

Walking behind her, Hollis heard the frown.

Joy pointed her to a large cubicle with a wooden desk and two metal chairs. Hollis took the chair in front of the desk.

"Joy, do you remember speaking with a Catherine Briscoe sometime in the last three months? She was writing a story about community non-profits." Hollis had practiced this question, hoping it sounded harmless.

"Catherine, Catherine .... No, I don't think so. I told you we don't use last names here. Even so, I don't remember a Catherine." Joy was sitting in a rocking chair that creaked loudly. Her smile was almost angelic.

Hollis took out a picture. "Here she is. Maybe she told you she was writing about Dorian Fields and his contributions? I'm working for the law firm that's defending Cathy—Catherine— and *Transformation Magazine*, the publication that hired her."

Joy's smile vanished.

She looked down at the picture but didn't take it from Hollis. "Mr. Fields is a good man. Without him Heaven's Praises wouldn't be here. Hell, without him *I* wouldn't be here."

"I see. So, do you remember talking to Catherine about Fields?"

Hollis noticed the changed attitude. Joy's right eye twitched and she kept rubbing her nose.

"Hell yes, I remember her, pushy bitch. She practically accused me of stealing. And I called Mr. Bartlett, too." Joy rocked furiously back and forth.

Hollis instinctively sat back. "I'm not accusing anyone, Joy. I just want to get a copy of your annual report for last year. Since you also receive government money you have to have one. May I have a copy?"

Joy stopped rocking. "Get the hell out of here. I don't have to give you anything."

"That's true, because I don't have a subpoena for your records. But I can come back with one, and if I do I'll ask for a lot more than your last annual report." Hollis looked her straight in her twitching eye.

Joy's fury rolled off her like heat waves. "Wait here."

"HOW DID IT go?" Mark said, a sandwich in one hand and a pen in the other.

Hollis briefed him on her visit to Heaven's Praises.

They had decided to meet for lunch at a little café down the street from Hollis' condo.

"Promise me, if it even looks like there is the slightest chance I might need homeless services, you will not take me there. That Joy is scary." She tossed the annual report on the table. "The thing is, it looks legit. Fields was by far their largest contributor in goods and cash. But I don't know enough about running homeless shelters to judge if there was something amiss. I need something to compare it to."

"Let me take it back to my firm. We have a great tax attorney; he can let us know if anything is there." He scribbled a few words on his pad. "What else?"

Hollis told him about the hidden window sill safe in Cathy's condo.

"So, there may or may not be something we could use?" Mark asked.

"Maybe, although on the face of it there was nothing in there worth the secrecy. But to know for sure, I'm going to take my time with each item. There has to be something there worth hiding."

"You know how much we need signs of a solid lead," he said. "Start working on it."

Hollis' brow furrowed. "Mark, did you know that Cathy personally helped a woman to get her life back on track?"

He nodded. "Not one woman—a women's shelter near Fruitvale Avenue. There were three women."

"How do you know?"

"Because she always used to hit me up for money." Mark smiled. "I take it you weren't approached."

Hollis shook her head. "No. She never asked me. I guess because she knew I didn't have any extra money."

"I was probably an easier touch." He paused. "I asked her once why she got involved. She was pretty cryptic and answered, 'Because I can do it.' "

"She was an interesting lady." Hollis shrugged. "Going through these papers and articles, I still haven't figured out if Cathy ever had anything that could nail Fields. But I can see why she was desperate to get my help to gather the rest. I just wish I knew what that comprised." She paused. "And yes, before you ask me, I told the police about the hiding place and gave them the originals."

"Good, and point well taken. I don't think she had enough to really sink Fields. But she must have thought she would have her hands on it soon; otherwise her behavior would have been uncharacteristically irresponsible."

"No, I reject that. Cathy would never have come to see me if she didn't know where she was going with the story. And she never would have contacted me if her article was bogus."

"Then all we have to do is prove it." Mark reached into his briefcase and pulled out a two-inch stack of papers. "Here's the last of Transformation's discovery. They gave us all the discovery files they requested from Dorian Fields. Fortunately he's legally required to respond. You stick with verifying the non-profit programs; I'll review the accounting records and see if there could be two sets of books."

"Good idea. I've got to wrap up the Koch matter. I'm getting signs from George that his patience is wearing thin. I need to finish going through those letters and decide if there are heirs we need to contact. It won't take me long; it's just that there is so much to juggle right now."

"I'm beginning to agree with George, how long does it take to read a few letters?"

"I don't want to just skim through them; I get the feeling that the payoff is there, but hidden somehow."

"We have only fifty-eight days left. Read the letters and move on. I need you to finish visiting and evaluating the community centers."

"Okay, okay. I'm visiting another non-profit tomorrow. In addition, I'll sort through the discovery and write up an analysis of contents from the window sill. I'll have it done by the end of the week." She hesitated, "But, Mark, those letters aren't easy to read."

He gave her a scrutinizing look. "What do you mean? Are they too graphic or emotional for you?"

"No, no of course not," she said, packing up her briefcase. But she could hear the lie in her own voice.

CHAPTER NINE

～～

WELCOME HEARTH INDUSTRIES was located in a run-down strip mall in the industrial section of San Lorenzo. Traffic sounds drowned out the ping of the bell over the door. Hollis looked over her shoulder, noting the small crowd of workers waiting patiently in line in the front of a food truck painted a garish yellow. Hollis put her hand to her eyes to block out the glare of the sun.

"Miss Morgan?"

A small, thin Asian woman opened the door into a large sitting area. Chairs were arranged classroom style, facing a whiteboard. She wore denim jeans and an un-tucked starched white blouse. Her hair was neatly done in cornrows and pulled into a ponytail.

"Yes," Hollis said. "Cynthia Lin?"

"That's me. Can I get you some water?" She pointed the way down a narrow hallway lined with pictures of smiling children and seniors.

They entered another good-sized room with the largest round conference table Hollis had ever seen. She took the seat closest to the door.

"Come in, did you have any trouble finding us?

"No, no trouble at all, and, yes, Miss Lin, some water would be very nice."

"Call me Cynthia." She went over to a shelving unit at the far end of the room where a small refrigerator appeared to be fully stocked with bottled water. She took out a bottle and handed it to Hollis.

"You must call me Hollis." She opened the bottle and took a deep swallow of the ice cold water.

Cynthia smiled and sat next to her.

"I represent a client who wrote an article for *Transformation* magazine that may have shown Dorian Fields in an unfavorable light. Mr. Fields filed suit against my client, but she was killed a few days ago and all her information has … has disappeared."

Cynthia said nothing but nodded her head in sympathy.

"Wade Bartlett called and told me the circumstances. He said you might be visiting. When you phoned, I knew who you were."

Hollis raised her eyebrow. They had gotten the names of centers to check from *Transformation*. Somehow Bartlett found out and called ahead to all their donees.

"I see, well, what I'd like to know is, how much do you count on Fields' support? Is he a major contributor?"

"Mr. Fields is very good to us. He pays for all our food and most of the clothing for our clients."

Hollis glanced at the pictures of smiling faces that covered the walls in this room as well.

"Cathy Briscoe noted that she spoke with you about three months ago. She had a question mark next to your name. Do you know why?"

Cynthia shook her head and shrugged.

"Tell me, what actually does Welcome Hearth do?"

Cynthia straightened in her chair. "What do you mean?"

Confused by the non-answer, Hollis looked down at the printout she had gotten from *Transformation*.

"I mean, is the group a homeless shelter, or a job locator, or ….?" She held her palms up in a human question mark.

"Oh, I see what you mean." Cynthia licked her lips. "We are a community resource."

Hollis bent her head down slightly, urging her to say more.

She continued, "Er … we give out food and we give out clothes …. Do you want to see our brochure?" Her hand shook as she rose from the chair.

Hollis' forehead creased. "Yes, sure."

Why the nervousness?

Cynthia left to get the brochure, and Hollis slipped out behind her. Three closed doors opened onto the hallway and at the end an office, where Cynthia could be heard opening and shutting file cabinet drawers.

Hollis got up from her chair and with one eye on Cynthia, took a peek inside the first door. It was a storage closet with six shelves. All but one was empty, and it only held a first-aid kit. Cynthia reappeared in the hallway.

"What are you doing?" She darted toward Hollis, putting her back to the closet door.

"I was trying to get a sense of Welcome Hearth." Hollis wasn't ready to back down. "Where are the …."—she looked down at the paper in her hand—"the five other employees?"

"Everyone is out doing errands." Cynthia didn't look her in the eye. "I'm here to answer the phone."

"Oh, what kind of errands are they doing?"

She mumbled what sounded like a curse word.

Ignoring her, Hollis reached inside her bag for her camera. "Can you show me around?"

"No, I can't." Cynthia's voice rose. "Mr. Bartlett said people are entitled to their privacy."

Hollis wanted to look behind the other doors, but it was clear that Cynthia was not going to let that happen. Fields gave them clearance to visit whatever locations they chose, but it wasn't a surprise that Bartlett had cued staff.

"Everything is so clean and quiet. Do the homeless sleep here?"

Cynthia didn't answer, but the set of her jaw spoke volumes.

Hollis decided not to push her luck. "Okay, well, I'll take the brochure and be going."

Cynthia thrust a thin multi-colored brochure toward Hollis, who took it and put it in her bag.

Walking a half-step from Hollis' heels, she steered her down the hall to the door. "Goodbye."

"Good—"

Hollis was cut off by the slam of the door.

"I LIKE YOUR new office, Mark," Hollis said, looking around the room. He had placed two leather maroon visitor chairs around a small table and another in front of his oversized walnut desk. Sand-colored drapes bordered large double windows. It was comfortable without being cozy. "Impressive."

"I won a very profitable corporate case." Mark shrugged. "It's not a corner office … yet, but it suits me."

"One day I'll have one." Hollis couldn't keep the wistfulness out of her voice.

"Has Triple D extended you a contingent offer pending your passing the bar?"

Hollis waved her hand back and forth. "Not formally, but I think they think there's an understanding."

"You seem hesitant. What's the problem?"

"No problem. I just don't want to have things end like they did with Cathy. Cathy was a star; she passed the bar on her first try. Her work at Triple D, while not stellar, was commendable. But still she became disillusioned and left to be a writer for a tabloid, so … go figure." Hollis frowned, trying to articulate the sense of caution she felt but unable to put her finger on the source. "But all this … this story she was after, I honestly don't know. Something isn't right."

"You're not Cathy. She was never the perfectionist you are,

for one thing. Her work was passable, but not ground breaking and—"

"We're not talking about the same thing. You don't have to put her down to convince me." Hollis gave him a reproving look. "I was aware of Cathy's shortcomings, but she had uncanny intuition. She couldn't always prove her case, but she was never wrong. But there's something I can't put my finger on that tells me she found out she didn't have the full story."

"Well, then, let's prove her right." Mark opened a file. "What happened at Welcome Hearth?"

Hollis recounted the meeting with Cynthia Lin.

"So we were right. Bartlett has called ahead to all the centers."

"And you should read the brochure." She passed it across the desk. "Other than six half-pages of glossy non-speak, we could be talking about a farm or a car repair shop."

Mark picked up the pamphlet, unfolding the copy from front to back.

"Did you get an annual report?"

"That's just it. This time a copy of the annual report was in our discovery. It didn't say much, except that roughly sixty percent of their funding goes to administration. That's a chunk of change. But you should have seen Cynthia's response when I asked about the five salaried employees."

Mark tapped his fingers on the desk. "What do they do?"

"I haven't a clue. Nobody was there. Cynthia said they were out on errands. Give me a break." Hollis closed her notebook. "I'm going to request payroll records and cross-check them with tax filings."

"Sounds like a plan, but that could take some time. Keep your focus on the non-profits. I'm doing background checks on Fields' executive team. Cathy highlighted several lines of notes about Fields of Giving's management. She might have found something."

"Staffing discrepancies and management questions. You know, Mark, I think we may have our first loose end."

"Let's see where it takes us." He put a sheaf of papers in a folder. "I've got to get going. Rena's cooking a special dinner."

"Yeah, I've got to get home, too." Hollis stood.

Mark looked at her. "Have you got plans? Would you like to come over? There's always plenty."

"Thank you, but no. Go home, Mark. That's where I'm headed, to my home."

YOU WOULD THINK that with 120 channels Hollis would be able to find something on TV that would hold her attention. Turning the TV to mute, she picked up a book she'd bought from her favorite used book store before beginning to study for the bar.

The ring of the phone caused her to do a time check.

"I hope I'm not calling too late." It was Brad Pierson.

A smile crept across Hollis' face. "No, I was just reading."

"Oh, you're a reader. Me too, but I just don't seem to find the time."

"I'm glad you said 'don't' and not 'can't.' "

He chuckled, "No, I realize that if I really wanted to read I'd make the time."

"I'm sure you still have redeeming qualities."

"Yes, I do as a matter of fact, and one of them is getting to the point." He took a breath. "I've got tickets to hear Yo-Yo Ma at the Pavilion Amphitheater next Sunday. Would you go with me?"

Yo-Yo Ma. Hollis mentally went through her closet, reviewing what she could possibly wear. Then she realized that she hadn't answered. "Yes, yes I'd like to go. I love him."

"Good, I thought I lost you there for a moment." He laughed. "I'll pick you up at five. You live close enough that it will give us plenty of time before the show to settle in and talk. If it's okay with you, we can have dinner at the amphitheater."

"Of course, if I remember correctly, the best seats for the show are the dinner seats."

Hollis was glad that Brad clearly didn't like chitchat. After she said she'd email him her address and directions, he quickly got off the phone.

Hollis held her knees to her chest and squeezed them tight, glad no one could see her silly grin. A few seconds later she felt a frown crease her forehead. How did Brad know how long it would take to get to the Pavilion from her house?

IT HAD TAKEN almost two weeks for Hollis to hear back from her friend who worked in the police forensics division about any information on Kelly Schaefer. The rental car plates and a not-so-great set of prints from a piece of notepaper weren't much to go on.

The police lobby was noisy with people waiting in a staggered line at the information window. Children ran restlessly among the adults, unaware of the undercurrent of tension that was the nature of the location. Hollis followed the signage to a small anteroom with a single door and an officer sitting at a desk in front of a computer monitor. She sat in one of the few plastic chairs that lined the wall on one side and took a deep breath, wrinkling her nose at the faint rank odor of sweat. Keeping her head turned away, she pretended not to notice the young boy sitting two seats down who alternated vigorously scratching his arms and slapping the side of his head. If the day officer could ignore him, so could she.

Five minutes later, she took a deep breath and approached the desk.

"Officer, I think that young man is in some kind of distress. Shouldn't you call for a doctor or—"

"He's a recovering addict on the other side of withdrawals. He's okay." The officer didn't look up from the computer screen.

"Oh." Hollis looked at the forlorn figure, bent over, scratching furiously. His thin arms were barely covered by a faded tee shirt. His brown hair was dirty and his pants were stained and torn at the hem. "Still, shouldn't he be taken somewhere?"

"He's waiting for his mom to get out. She should be coming out tomorrow." He finally looked up at her. "I know it looks bad, but he's been here before—picking up his mom, I mean. His name is Vince. If I kick him out, he'll go looking for a fix. He wants to go clean. I let him hang around until the shelters open up for meals and beds. He's harmless. He's been here every day since she's been in. Now, why are you here?"

She looked down at her watch and compared it to the clock on the wall. "I'm waiting for a friend, Stephanie Ross. She knows I'm here. She said she might be running a little late."

He nodded. "Sorry I can't offer you a better seat. Don't worry about Vince. You should have seen him when he was hooked. His mom has been in and out of jail for months, and he's always taken care of her. Too bad she never felt the same way about him."

Hollis walked back to her seat. She could see the bones in Vince's elbows, and the sunburn on feet that appeared to be a size smaller than his battered tennis shoes.

"Are you hungry?" She leaned toward him.

Vince abruptly stopped scratching but didn't look up.

Hollis reached inside her purse. "Here's five dollars … for you … to get something to eat."

Vince held his head up. Hollis was startled by the intelligence in the beautiful brown eyes that looked back at her.

"No, no don't give me money," he said. "Can … can you buy the food for me?"

His raspy voice sounded deep for a youth.

"Oh … ah … sure. I think there's a store on the corner. What do—"

Vince groaned, slapped his head and went back to scratching.

Hollis went over to the officer and explained that she would be right back in the event Stephanie came looking for her. Spanning the block in minutes, she quickly picked up items at a Seven-Eleven: a banana, a ham sandwich and a small carton of milk.

When she returned ten minutes later, the scene was unchanged, except for the uniformed EMT sitting sat next to Vince.

"Look man, you're doin' good. Don't give up." The burly African-American man leaned over to meet Vince's gaze. " 'Cept man, you can't stay here. Jim lets you stay here on his shift, but he's goin' off in a little while. He called me to take you to the center. You can come back tomorrow and pick up your mom."

Vince looked panicked.

"No … no I can't make it there. Too many people usin'. I'll go looking for some stuff." He looked up and saw Hollis standing with the bag in her hand. He reached out.

The EMT noticed her too and moved aside as she handed the food to Vince.

"Thank you," Vince said, even as he pushed the bread in his mouth. "See, Joel, I'm eating. I'm hungry, and I'm keeping it down."

"That's good, man." The EMT stood and held out his hand to Hollis. "He can't handle the milk." He handed the carton back to her.

"Oh, sorry," she said and slipped the container into her purse.

"You know Vince? I'm Joel Cannon. You a social worker?"

"No, I'm waiting for a friend. I just met Vince …. Well, not 'met' really. Anyway, I thought he could use a meal."

The side door opened and Hollis acknowledged Stephanie with a smile as she entered the room. Dressed in her lab coat and carrying a Dooney & Bourke bag, Stephanie quickly walked over.

She gave Hollis a hug and an air kiss on the cheek. Joel and a chewing Vince looked on.

"Hi, luv, what's going on?" Stephanie glanced at Officer Jim, who just shrugged and went back to his computer screen.

"Stephanie, this is Joel and this is Vince." Hollis pointed to each. "Vince has … has a problem, but he wants to be here when his mother is released tomorrow."

Stephanie looked at each of them and without a word bent over Vince and looked him in the eye. He didn't pull back from her scrutiny. Hollis saw her take in the tracks on his arm.

Hollis put her hand on her arm. "Stephanie, can I talk to you for a minute?"

She steered her away to the opposite corner of the room.

"Ah, no, Hollis, no, no. I was looking forward to lunch and I have that information you wanted about Kelly Schaefer." Stephanie followed her to a far corner.

Hollis put her finger to her lips. "Keep your voice down. We'll have lunch, don't worry. But isn't there somebody or someplace that could help Vince?"

"Who in the hell is Vince?" Stephanie threw up her arms. "No, wait, I can tell you. He's a junkie in bad shape. I thought you wanted me to help you with your friend who was killed."

"Shhh, not so loud, Steph. I do need your help with a couple of Cathy's items. But Vince is close to getting clean." Hollis didn't know if that was true, but she felt it was. "He just needs a chance. I don't know why his mother is inside, but the fact that he cares for her says he's not too far gone."

Hollis gave her the biggest pleading look she could muster.

"Since when did you start to collect abandoned puppies?" Stephanie crossed her arms over her chest. "All right, all right, at least he's able to eat. There's a non-denominational church, The Eternal Lantern, about a half-mile from here. I'll call the director."

Hollis took a deep breath. "Thanks, Stephanie. We can drive him over on our way to lunch."

"No, we can't. I'm not getting in a car with him." Stephanie smoothed her belt. "Tell—what's his name?—Joel to take him." She walked back through the double doors to make her call.

Judging from their looks, Hollis realized that Joel and Vince did not have high hopes from the conversation. They both straightened as she walked over.

"There's a church in the neighborhood that can help you Vince, the Eternal Lantern. Stephanie is going to see if they have room for you." Hollis couldn't ignore his distressed look. "You can stay the night, maybe longer."

Joel smiled at Vince. "It's a good place. Hard to get into 'cause they don't make you listen to the preachin'. People like it better. If you can go there, Vince, you'll like it. They'll take care of you."

Vince looked doubtful. "You sure I don't have to hear a sermon?"

Joel was about to answer when Stephanie returned with her fingers forming an "okay."

Joel tapped him on his shoulder, "Man, you better git up and git goin'. You're gittin' a break."

Vince rose, a little unsteady. He faced Hollis, lightly scratching his arm. "What's your name? Why did you help me?"

"Hollis Morgan." She tried to find the right words, "Because I could."

LUNCH WAS DELICIOUS. Hollis and Stephanie ate at the same bistro whenever they went out to lunch. Stephanie was the original poor little rich girl turned poor girl. Her father had been caught giving insider information in a Wall Street broker sting, forcing his family to turn their backs on their lush lifestyle. They had met at Triple D after Stephanie had just earned her degree in Criminology, when she was still adjusting to her lower income bracket. They had struck up a friendship immediately. As a crime lab assistant, she had the best City contacts.

"I owe you," Hollis said, scooping up her last bit of tiramisu.

"Yes, you do, and I'm going to collect someday, so don't forget." Stephanie dabbed her mouth with her napkin. "I hope you realize the risk I took in tracking down this license number."

"I thought you could do it in the course of your job," Hollis said. "I don't want to get you in trouble, Steph."

Stephanie waved her hand. "Don't worry about it. I'll add it to your friendship tab." She reached into her purse and pulled out a letter-sized envelope. "Here's the info on your girl. What's the interest? There are no wants or warrants."

"Just checking, I didn't hold out much hope that you would find something." Hollis quickly scanned the single piece of paper.

"I know I don't have to ask for what purpose you wanted this information."

"Don't worry. It's for a probate matter I've been assigned." Hollis eyed the sheet. "It says she's twenty-five years old; I thought she might be older. Schaefer is her married name."

"You mean I did this for your fancy law firm? That does it; you can pay for lunch." Stephanie put on her sunglasses.

"Not a problem." Hollis patted her case. "It was worth it."

ON THE WAY to her office, Hollis looked down at the message Tiffany had handed her. Hollis gave her an acknowledging smile. She was Triple D's longest employed receptionist. She had lasted almost a year, maybe she might make it.

The message was from *Transformation* magazine. She punched in the number and Devi's secretary answered.

"Miss Morgan, Mr. Devi wanted to let you know that Miss Briscoe left behind a few file folders that were with our attorneys. They could be duplicates, but if you'd like them, I can leave them at the *Transformation* front desk for you to pick up."

"I'll be there later this afternoon."

Hollis leaned back in her chair and rubbed her forehead. The way Cathy's files kept popping up, by the time she and Mark collected them all, they just might have enough ammo to fight Fields. She picked up the phone.

Mark's voice didn't hide his enthusiasm. "That could be great news. Let's go over where we are this evening. I can't stay long. I've got another case I'm working."

"Not a problem, why don't we just get together tomorrow."

"Can't. I'm flying up to Portland tomorrow for two days. I want to make sure we get all our depositions completed over the next week. We have a thirty day window. I made an interesting discovery; Cathy's notes point to a board director she wanted to depose as a last resort."

"She wrote that—'a last resort'? What does that mean?"

She could hear Mark rifling through pages. "In her notes— the ones you found hidden in her condo—Cathy indicated that one of Field's former board directors resigned abruptly." He paused. "Listen to this: evidently she interviewed him for almost three hours and for whatever reason promised him he would only be deposed as a last resort."

"Okay, I'll meet you in your office at six, and bring those notes with you. We need to consolidate every file and folder we've collected from all sources."

CHAPTER TEN

THE CONFERENCE ROOM floor was strewn with papers that could no longer fit on top of the paper-covered table. Hollis glanced at Mark, punching numbers into his cellphone.

They were in the evening quiet of Mark's office. Hollis wasn't able to commandeer a Triple D conference room, and while her condo was in a more convenient location, she preferred to keep her private life private. She had moved to the new condo in San Lucian to move past the memory of the terrifying confrontation with her ex-husband's killer. That, and the fact that the homeowner's association in the community she previously lived in made it clear that if she continued to have regular visits from law enforcement, she might want to try a neighborhood with different standards.

She could hear Mark trying to convince Rena he would be on his way home in a few minutes. After a moment, he put the phone back in his pocket.

"Okay, where are we? What do we know?"

Hollis reached over and picked up a pad of paper. "Cathy visited all of the Fields' charity centers in California. *Transformation* indicates that six of those centers are in the

Bay Area. I've been to two, neither of which appeared to be what they were supposed to be. But to be honest with you, while there was a lot of smoke, I couldn't find any fire. If Fields is faking philanthropy, he's going to an awful lot of trouble."

Mark tapped his writing pad. "So, Cathy writes an article that blasts Fields. And our charge is to find the proof—not the hint, not the smoke, but the solid proof that the guy is a fraud. There's something we're missing; you've got to go to more centers."

"Mark, suppose it was something else altogether? Suppose Cathy stumbled onto another, bigger story?"

"Forget something bigger. We don't even have a handle on something small. We need to substantiate this article." Mark flipped through pages. "You've got to check out at least one more non-profit. Try this one." He handed over a sheet of paper.

She sighed. "What's one more? I'm on a roll." She quickly read through the text. "Open Wings. What is it, a homeless shelter?"

"I can't tell. That piece of paper is the only thing we have on them. Notice how much they get from Fields—a million a year."

"Wow, that's a lot of money. It must be the Rolls Royce of homeless shelters."

"Can't wait to hear what you uncover." He started packing his briefcase. "I'm bushed. I have an early flight into Portland again. I'll be returning late tomorrow. Let's talk again Wednesday."

Hollis picked up the papers and sorted them in an expandable folder.

"If it's okay with you, I'll make arrangements to visit the shelter later in the week. I need the next couple of days to work on another client matter."

"Is George concerned about the time you're giving to Cathy's case?"

She shrugged. "He's fine with it, as long as I finish reading the Koch family letters and determine whether or not there's an unknown heir."

Mark locked his briefcase. "What's the problem? It's not like you to drag things out. Finish the letters. Remember, I start taking the non-profits' depositions week after next, and I need your discovery summaries as soon as possible."

"Don't worry, Mark." Hollis muttered. "You'll have your summaries."

"Thanks," he sighed. "I know you don't need to be pressured, but we've got to be ready to fight Fields' claims."

She walked out as he turned to lock the door.

"No pressure needed, I'm on it." Hollis bristled then she paused. "Mark, Cathy's libel suit is just a small part of a deadly picture. You remember that old rhyme: 'sticks and stones may break my bones, but words will never harm me'?"

He looked at her curiously. "Yeah?"

"Well, it's not true."

EARLY THE NEXT morning, Hollis sat behind her closed office door. Her work was divided into two large stacks on top of the lateral drawer cabinet. To her right were files and folders relating to Cathy's case; to her left, Margaret Koch's box of letters and a pad for notes.

She opened the letter from 1947 at the top of the stack.

> Dear Maggie,
>
>   I'm not going to judge you. What good would it do? You are family. But you need to know that Charles has started drinking real bad. When he heard you married Eric, he changed. We think it best you don't come back to Rowan for a while. I know it's been two years, but a lot of people here liked the Ferris'. They helped a lot of families.
>
>   Now people here feel sorry for Charles losing his arm and leg like that in the war—then you left him, too.

It wasn't good what you did.

> Your cousin,
> Lisbeth

Hollis sat back in her chair, her brows drawn together in a scowl. She pulled out a small wrinkled piece of crumpled stationary that had been smoothed and tucked into the corner of the envelope.

Margaret,

I knew you would come back. Meet me at the old leather factory at ten o'clock. Carry a flashlight, it can be dark out there.

C

Hollis grimaced. Margaret went back to Rowan. Why? She flipped the paper over, but there weren't any markings identifying the sender. There was no doubt in her mind that "C" stood for Charles. She picked up the next letter.

Dear Mrs. Ferris,

I am sorry to inform you that your husband's appeal was denied. The original sentence of twenty-five years will stand. The state will move him from Cook County Jail to Joliet Prison on Friday, August 31, 1948. We have explained to your husband that with good behavior he might be released in fifteen years.

You will be allowed to visit him for one hour on Thursday, August 30, 1948, at 9:00 a.m. If you would like me or Clive Campbell to accompany you, please let me know.

> Sincerely,
> Michael Dyer, Attorney
> at Law

Hollis thought she was past someone taking her totally by surprise, but Margaret continued to surprise her.

> My Darling,
>   I'm sorry you weren't able to make it to visitor's day. I thought maybe this time they would call that I had a visitor. I don't understand why I haven't heard from you. Life here is bearable, but just barely. The only thing that keeps me going are your last words saying you would wait for me. I need to see you, Margaret. If I can't touch you I need to see you.
>   I can't believe it's been two years since I saw you run away from Charles' body. You didn't see me. But I saw you standing there on the ridge, looking down on him. There I was with the two people I loved most in the world. We never talked about that night, and I will never mention it again. But inside here it's all I ever think about. I have no regrets, and I forgive you for what you did. My brother had changed, he was so full of anger and resentment I didn't recognize him. I would serve two life sentences for you, my love. But I need to hear from you. Please visit, or write.
>
>   Your loving husband,
>   Eric

Shaking her head, Hollis reached for the bottle of water on her desk. Margaret was leaving a bad taste in her mouth.

She pulled out the next envelope, only it wasn't a letter. It contained a legal notice. It was a statement of Dissolution of Marriage, dated 1951. Margaret had divorced Eric in jail after three years, apparently not long after she had gotten the letter that Hollis had just read.

She felt a chill run along both her arms. She had divorced her own ex in prison, except in her case he was the one who should have been there. She took another swallow of water.

She picked up a letter, dated 1951.

Dear Margaret,

I'm sorry it took so long to get back to you but you sent your letter to the wrong address. I retired from the library and moved away from Rowan some months ago. I missed you at Lisbeth's funeral, but I understood you were on your honeymoon. You may not believe me but I do hope you find happiness with Michael Koch.

To answer your question, Eric settled in California. When he got out he didn't want to have anything to do with Rowan. I guess some people never thought he should be let out. I guess I was one of those people. I hate to tell you these things, but you said you wanted to know. I guess you can close the book on this part of your life.

Regards,
Mary Ellen

Hollis scanned through the box for the remaining letters. She checked the dates. They were all written within twelve months of each other.

She hesitated to reach for the next opened letter. Then, with a sigh of resignation, she lifted out the page of lined pink stationery. She might have imagined it, but a momentary rose fragrance drifted up.

She put the letter back unread and picked up her phone.

"Stephanie, how about a … a movie tonight?" Hollis spoke quickly into the phone before realizing she had reached an answering service. After saying she didn't want to leave a message and agreeing to have a nice day, she hung up.

She pushed another number.

This time Hollis got to the point of her call quickly.

Mark replied, "Gosh, Hollis, I'm beat. I just got in from the airport. Ordinarily we'd like to get together, but this is a weeknight and I've got a crazy day at the office tomorrow."

She mentally kicked herself. "Of course, I forgot you were out of town. How was Portland?"

"Not bad. We'll see how negotiations go."

"Well ... good."

"You okay being by yourself?"

"Sure, sure. I was trying to think of an excuse to keep from having to finish organizing all my study papers. Looks like tonight will be the night."

"Are we still on for tomorrow?"

"Of course, actually I need to polish the summary and make up a digest of all the notes and reports, so we don't have to waste time going through papers."

Mark paused. "You could just relax, you know."

A quick retort came to Hollis' lips but she held it back. "I'll see you tomorrow."

Driving home, she reflected that her desire to read the letters slowly was more than just prurient curiosity about the lives of others. On some level, she marveled at Margaret's self-centeredness and her ability to walk away from the lives she left in a tumble.

She was gaining insight into the life of someone who wreaked havoc, instead of the one who had havoc wrecked upon her—namely, herself.

IN THE OFFICE the next day Hollis, after a night of reasoning with herself, was ready to quit stalling and get back into George's good graces. She slid the information Stephanie located for her from the envelope and dialed Kelly Schaefer's phone number.

"Hello." The voice sounded distracted.

"Kelly Schaefer? This is Hollis Morgan. We met at Margaret Koch's home about two weeks ago."

There was silence.

"How did you—"

"I got your phone number ... from public records." Hollis rushed her words. "I know you said you were looking for something when we met at the house, and I think I have what you were trying to find."

"What have you found?"

Hollis paused, and then made a decision. "I have letters—letters to Margaret Koch."

It was Kelly's turn to pause. "How many are there?"

"Quite a few."

Not sure why, Hollis decided only to tell her about the letters, not the clippings, and only the letters that referenced Rowan.

"If I show you my letter will you let me see the ones you're holding?"

"You have just one letter?" Hollis pulled out a pad and scribbled a note.

"Yes."

"When can we meet?" Hollis called up her calendar on her computer screen. She looked up to see George standing in the doorway. She pointed at him to sit while she waited for Kelly to pick a date.

"Got it." She wrote on her calendar. "See you then."

"HOW DID YOU track her down?" George scanned the notepaper Hollis passed to him.

"It wasn't easy. I think there was a public record." She pretended to look through the file on her lap. She wasn't ready to reveal Stephanie as her secret source. "But George, I noticed there was nothing in the detective's file about Koch's second husband, Eric Ferris, let alone his criminal record."

George nodded. "I know it's curious, but we told him to look for Margaret's heirs, not to prepare an exposé. Do you think this Kelly person is really an heir?"

"I don't see how. I still can't find any blood relationships, but it's possible."

"I'm going to give Brad Pierson a call. His firm is going to owe us a refund."

Hollis raised her head in protest. "No, George, wait. Let me find out exactly who Kelly is and finish reading the letters. We don't know anything for sure. I want to follow up on a few things that Pierson may have overlooked."

She took note of her own lie.

CHAPTER ELEVEN

***

HOLLIS WAS READY to go home, but there was one more thing she wanted to do before leaving the office. She had put it off for as long as she could. She knew intellectually it was too soon to search online for the pass list from the California Bar, but she couldn't help herself. She clicked on the website—nothing. Historically, bar exam results weren't released for another four to six weeks. At this rate she would be mentally certifiable in two.

She shut down the computer and headed for home.

Walking up the path to her condo, she returned the buoyant greeting from her next-door neighbor with a quick, dismissive wave. Fortunately she wasn't counting on being elected president of the Neighborhood Watch Association. Reaching inside her curbside mailbox, she withdrew a handful of bills and advertisements.

Inside the front door, she kicked off her shoes and tossed her purse on the entry table, at the same time pushing it back into place against the wall. Still, it wasn't until the magnitude of the mess—the overturned chairs, emptied bookshelves and ransacked furniture—caught her attention that her situation sank in. With a sickening ache in her stomach she slid down the wall and crouched on the floor. She rummaged around in her purse and found her cellphone.

A calm male voice answered, "Nine-one-one dispatch."

Hollis took a breath. "Can you send the police? I've been burglarized."

"Are you alone in the house?"

"I-I don't know. I think so."

"The police have been alerted. They notified me that unless you're under an immediate threat, they'll be there within the hour. Would you like me to stay on the line with you until they arrive?"

Hollis thought about it a moment but then reassured him—and herself—that she would be fine.

She walked around the living room, careful not to touch anything except for the small frog sculpture that lay in pieces on the floor. It had belonged to her grandmother. She placed the fragments carefully on the counter. The kitchen wasn't spared; several drawers had been pulled out but not emptied. She sighed in dismay when she touched what was left of her battered laptop on the table.

She went upstairs to peer into the remaining rooms. They had left her roomful of boxes and junk alone. Only her bedroom had been tossed. All the contents of her dresser were dumped on the bed; ditto with the contents of her closet. She had just installed closet organizers to finally bring order to the chaos. Now the chaos was back, tenfold.

She returned downstairs to the kitchen, put her head in her hands and sat down to wait.

Less than an hour later an unmarked car pulled up behind a police vehicle. Hollis could only imagine what her new neighbors were saying. She went to open the door.

"Hello, Ms. Morgan."

Hollis stepped back in surprise and then felt a smile creep over her face.

"Why, Detective Faber, since when does homicide show up for a San Lucian burglary?" Hollis moved aside so he and the female officer behind him could enter.

He motioned to the officer. "Officer Vega is here to get your information. I just heard your address come across the radio. I was in the area and I thought—"

"You thought you would drop by," Hollis said. She glanced around the room. "Well, I can assure you I'm usually a much better housekeeper than this."

He stepped into the room and gave her a sympathetic look.

Officer Vega had already slipped on blue disposable gloves and was walking around with a small notebook. She stopped in front of the large desk in the dining room that had been emptied.

"How long were you out of the house?" she asked.

Hollis sighed. "I just got home from work. I was gone most of the day. I actually came home early."

Vega looked into the kitchen. "You'll need to make a list of anything you find missing and get it to us as soon as you can."

Faber walked through the condo, went upstairs and returned to the living room. "I'm not convinced this was a burglary. I think they were looking for something in particular."

Vega's eyebrow lifted.

Hollis was taken aback. "Why would you say that?"

He pointed to the floor. "Only papers, books, and folders are thrown around. All your electronics are still here. The stuff that's easy for a real burglar to fence." He walked into the kitchen and used a pencil to poke at her computer. "Why take the time to destroy your laptop?"

Vega stood in the hall doorway. "It would help if we could take it with us. Was there something important on it?"

Hollis shook her head. "Not really, mostly just class downloads and my study pages for the bar exam."

Vega picked up the laptop and put it in a large folded plastic bag she pulled from her back pocket. "I'm going to head back to the precinct and write up a report. You can check online in forty-eight hours. You'll need it for insurance purposes." She handed Hollis her business card.

"Officer, I have a few more questions for Ms. Morgan," Faber said. "I'll meet you back at the station."

Vega nodded and left.

Faber frowned. "Ms. Morgan, do you have any idea who might have done this?"

"Could you please call me Hollis?" She righted one of the dining room chairs and sat. "No, I don't know anyone who would care enough about my law school notes and monthly bills to break into my home."

"Is there a friend you can stay with or someone who can stay with you?"

Hollis shook her head. "No, there's just me. I'll be okay." She sat on her shaking hand.

If Faber noticed, he didn't say anything. He rubbed his hand over his head. "Okay, all right. Vega will take it from here." He walked over to the front door. "Actually, there's another reason I came in on this. I … I bought these tickets to see Yo-Yo Ma on Sunday. I know it's short notice, but I just got them today and the only person I could think of to ask to go with me was you."

Hollis looked up to the ceiling.

"Detec—"

"John, call me John." He laughed.

His laugh caught her off guard. It was surprisingly contagious.

"John, I adore Yo-Yo Ma, and I would really like to go with you—"

"But."

Hollis nodded. "But someone else already asked me."

"Already asked you to hear Yo-Yo Ma?"

"I know. What a coincidence!" Hollis slapped her thigh. "I don't have a date for six … never mind." She blushed.

John laughed. "Look, maybe some other time then."

"I would really, really like that. Please ask me again." Hollis said in what she hoped was her most earnest voice.

He looked her in the eyes and said, "You can count on it."

After the detective was gone, Hollis just sat, dejected, in the middle of the room. Taking a breath, she started to put the sofa pillows back in place. She didn't usually cry, but she felt tears were just a few eye-blinks away. Pulling her thick hair back into a ponytail, she began to return books and CDs to their places on the shelves. Who would want to burgle her? She exhaled a long sigh and replaced the dining room chairs around the table.

It took the rest of the evening to return her condo to a semblance of order. She wiped her kitchen counters down with disinfectant and vacuumed the carpets twice.

A folded piece of paper under one of the corners of the living room throw rug caught her eye. It looked like binder paper—the type a child would use in school. It was folded several times until it was about one-inch square. She opened it carefully. She didn't watch much TV, but enough to know there might be fingerprints.

She froze. It contained her name and address.

John was right, this wasn't a random crime. She had something someone wanted, but what?

THE NEXT MORNING she parked her car in the rear lot of Open Wings. On one hand, she was reluctant to leave her home so soon after it was ravaged, but she knew she had to push past her fear and not let it hold her hostage. Cathy's case was more important.

Determined to get some real answers, she took a deep breath and entered the double glass doors. By now Hollis was used to visiting shelters. Open Wings resembled a store front with large corner windows and a modest sized reception area. A large sign in the window boasted a notary and clinic services.

Hollis was greeted almost exactly the same as in the previous two places. This director, who introduced herself as Lilia Martini, led her to a small office with a window looking out onto the reception area.

Sitting at a black metal desk with matching chairs, Lilia said, "No, Miss Morgan, I haven't received a call from Mr. Bartlett about you. But then I've been on vacation until yesterday." Lilia blew into a tissue, leaving her nose strawberry red. "But there's no problem. From time to time we get people wanting to see our facility for themselves."

Hollis looked out the window. "Is this your entire facility?"

"Yes. All the lockers are along that back wall."

"Lockers?" Hollis took note of the triple rows of beige lockers. It reminded her of the girls' gym in high school.

"For the mail."

Hollis flipped though her notepad. "Miss Martini, I'm sorry. I should have asked this when I first came in. What does Open Wings do?"

"We are a mail location for the homeless. They don't have permanent homes like you and I do. They need a place to collect their social security or disability payments. We give them an address they can use free of charge. By the way, you can call me Lilia."

Hollis smiled and nodded in understanding. She reached for the annual report and turned to the page she remembered reading.

"Lilia, how many employees do you have?"

"Counting me, one." She laughed.

It was starting to make sense, or at least a thread was beginning to form. She opened up the report for Lilia to see. "I'm a little confused. Did you prepare this annual report?"

"Me, oh no."

"Isn't this your signature under 'Director'? It says you prepared the report."

Lilia started to pale and licked her lips. "Yes and no. I signed it because this young lady from Mr. Fields' office told me to.

I'm not good at writing and arithmetic. Mr. Fields' office sent me this lady—I think her name was Phyllis—to help me do our reports."

"Do you know what it says?"

"Well, I guess I'm not good at reading, either. Mr. Bartlett said I could leave everything up to Phyllis."

Lilia's revelation told Hollis what she needed to know. Hollis wasn't going to ask about the missing four employees.

She made a couple of notes. "How much money do you receive from Fields of Giving each year?"

"Well I don't know, enough to pay me and pay the rent. Phyllis gets all the bills." She hesitated. "Is there a problem?"

Problem? Hollis thought to herself, not a problem except that the million dollars showing as total operating costs might be a tad overstated.

"SO HOW DID you leave it with her?" Mark had started to wear glasses, which were currently perched on the top of his head. "You did tell her you were representing a defendant?"

They were sitting in Triple D's meeting room with Cathy's files.

"I told her before I left and she didn't seem to mind," Hollis said. "I got the feeling that as far as she was concerned, since she was telling the truth, she had nothing to hide. Although, I bet if Wade Bartlett had gotten to her it would be a completely different story."

Mark slipped his glasses back on. "Okay, here's what we've got. The first hearing date is in two weeks, which is just a formality to go on record with the continuance Fields gave us. I should finish deposing Fields' staff next week. I'll have to work analyzing the responses into my schedule. What have you got to report?"

"My, my, you seem to be taking on the role of authority. What else have I got to report?" Hollis mocked, and then her smile faded. "I've got a lot of bread crumbs but no loaf. I do have this kernel of an idea. Cathy wrote that Fields was using his non-profits to launder money, but I wonder if there was something more …. We've got grossly inflated annual reports and barely operational charity organizations."

Mark ran his fingers through his hair. "I would give anything to know what Cathy found that would push her to risk everything on this story."

"The board director you mentioned, the one who wanted to be the last resort, can you identify him?"

"So far I haven't had any luck, but I'm not giving up. Cathy was pretty secretive." He leaned back into the sofa and rubbed his eyes with his thumb and forefinger. "You know money laundering is not so farfetched. It explains Cathy's focus. The problem is, we don't have a lot of time to prove it. On the other hand, if we stumble onto pay dirt, we are going to have some powerful people trying to shut us up."

"Yeah, uh, there's one thing I have to tell you," Hollis said sheepishly. "My condo was broken into and searched."

"What? When?" He looked around the room as if the burglary was taking place right then.

Hollis explained about the break-in and the police response.

"The thing is, I didn't have anything of real value in the condo. I had all Cathy's papers with me."

Mark shook his head. "This is getting out of hand. Cathy was murdered and whoever did it might not hesitate to kill again. Maybe we should let the police take over, or just let the whole thing drop."

"No way," she said. "Cathy died for her story. She touched a nerve, obviously a big one. We can't let Fields scare us away."

"He may not stop at trying to scare us. Hollis, you did tell the police about a possible connection between your burglary and Cathy, didn't you?"

Hollis couldn't look him in the eye. "Yes and no. No, I haven't yet and yes, I'm going to." She held up her hand to forestall his objections. "Honestly, Mark, I just thought of it myself.

Besides, John Faber showed up. He's aware of both cases. I'm sure he's zeroed in on the possible connection."

"This whole matter is taking on a new angle," he said. "Let me know what the police say."

"Wait, I just thought of something." She stood and started to pace back and forth. "Fields is not going to really come after us, I mean *directly*, because he'd be an obvious suspect. He won't want anyone else peering into his affairs. If anything, we're going to be protected."

Mark gave a small laugh. "Who said every cloud doesn't have a silver lining?"

HOLLIS PULLED OUT her cellphone to check for messages. She was pleased to see Gail Baylor had not called to cancel. Their meeting was still a go. She found the phone number for a Gail Baylor written in the corner of one of Cathy's pages of notes. It didn't take her long to track her down. It had been Triple D's receptionist, Tiffany, who gave her the idea that Cathy had to have someone at *Transformation* who helped her with the administrative side of her research. She had to have an assistant if only to handle logistics. Finally, after some minor poking around, she discovered that Baylor had been Cathy's assistant. Hollis frowned; it bothered her that Devi neglected to offer that piece of information, but it probably didn't occur to him to consider the clerical support.

On the phone Gail Baylor seemed nervous but affable and more than willing to talk about her last assignment. They agreed to meet in the lobby of the Oakland Library that afternoon; it was public and private at the same time.

Two minutes after their agreed time, Hollis looked around the lobby and spotted an older woman who was clearly waiting for someone. She was stocky, with graying blond hair styled in a chin length pageboy and black rimmed glasses perched on an overlarge nose. Her gray eyes finally landed on Hollis and she walked tentatively toward her. Hollis met her halfway.

"Gail Baylor? Hollis Morgan." She said, holding out her hand.

The woman shook her hand and flashed a smile that disappeared as quickly as it appeared. She nodded. Hollis pointed to two overstuffed upholstered chairs in an alcove away from the door.

"I'm sorry I'm late. I'm always losing track of time. Sometimes it gets me in a world of trouble. I hope I'm not making you late for your next meeting."

"No, no, not at all, please … Gail, you don't have to apologize. I just appreciate your meeting with me." Hollis smiled. "Had you known Cathy long?"

She sat primly with her hands in her lap. "From her very first day at *Transformation*, Mr. Devi assigned me to her. I get all the new ones." Gail leaned forward. "Don't get me wrong. I'm not complaining. You're going to think I'm being negative about the magazine. My supervisor wrote me up about that on my last evaluation. I should keep my mouth shut."

Hollis touched her lightly on the hand. "No, no, I don't think anything of the sort. I need your help. Please, tell me about your work with Cathy."

"Cathy was easy to work with. She treated me like an equal." Gail tucked a lock of her hair behind her ear. "She would let me edit her pages and even do some background research. Don't think I'm talking bad about the others, I'm not. Cathy just had a lot of work, and she felt I could help her."

"Did she explain what she was looking for? When was the last time she discussed her work with you?"

Gail's eyes grew large. "I didn't mean to mislead you. I assure you that the work was all hers. I was responsible for transcription. That's all. Talking too much is what gets me in trouble."

Her hands started to shake.

"Gail, it's okay. I've worked with Cathy, too. She loved to share ideas about a case." Hollis kept her voice low and calming. "She would talk out loud about what approach she planned on taking."

Baylor heaved an audible sigh of relief. "Oh, I see. Yes, yes, she would talk with me. That's how she made me feel special. We were a team. Management thought I could only do clerical work, but Cathy let me do research. I can be really obsessive; you know, making copies of copies. I would offer suggestions. But I don't want you to think she didn't write ev—"

"Gail, trust me, I understand." Hollis couldn't stop herself from interrupting. "What kind of research?"

Gail's head bobbed in a fast nod. "Okay, yes, of course. I'm sorry I get so distracted and I then I lose my train of thought. Once ...." she stopped when she saw Hollis' expression. "Sorry. I collected all the annual reports and then she had me pull what tax records I could locate. There weren't many, but she seemed pleased."

"How did she verify her facts? Where did she keep her fact finding?"

"I scanned everything and put it on a computer memory stick."

"Where is it now? There was nothing with her things." Hollis felt guilty hearing the impatience in her own voice and the increasing contriteness in Gail's.

"I guess the police have it. It had to have been with Cathy's things." Gail bit her bottom lip. "I want to say that Cathy was one of the most ethical people I knew. If she thought that Fields was crooked then she knew what she was talking about."

Hollis struggled to phrase her next question. "I can see why Cathy wanted to work with you." Hollis spoke slowly. "Do you know if Cathy got any threats, anyone who hassled her?"

This time Gail didn't have a quick response. She looked pensive and turned to gaze out the library's glass doors.

"We were finalizing the Fields story. Cathy was happy it was completed and into legal for review. She was out of the office and I was clearing her desk when a call was put through." She squinted with remembrance. "I wasn't being nosy. It was just an accident I was even th—sorry—it was a man who wouldn't leave his name but did want me to take a message for her."

Hollis tilted her head. "What did he say?"

"Back off."

CHAPTER TWELVE

H OLLIS LIFTED THE Post-it off her computer screen. She had worked a full day yesterday, so the message must have been left late last night.

> Let's meet at 3:30, after you've finished reading the letters.
> —George.

Hollis crushed the note and threw it into the trash can.

Grabbing her mug and a tea bag, she went into the lunch room for boiling water. On her way back she stopped to refill the water cooler, and when she could think of no further reason to delay, she returned to her desk.

She took out the box. Only three letters were left and two were stamped "Return to Sender."

Hollis opened the earliest "return to sender" letter. It was written in 1955.

> Lisbeth,
>   I'm glad you've befriended Eric. From what I've heard, he needs a friend. I dislike putting this in a letter, but you

refuse to take my calls. Even though we are cousins, I could also use a friend.

Lisbeth, I did not kill Charles.

Despite what Eric may have told you, he killed Charles. There, it's out. I saw him that night standing on the ridge after he had left Charles' body in the ravine. I think he knew I saw him but we silently agreed never to talk about it. Even when the police came and took him away, and he professed his innocence, I never said a word.

I may not have kept in touch with you over the months, but you didn't reach out to me, either. I could have used a family member's shoulder, or a friend's hug to carry me through those dark days. I couldn't face Eric after he went to prison; I knew my disappointment and regret would show in my face. I didn't want to see him. Your judgment of me reached across the miles. Yes, I turned to Michael Koch; he gave me kindness when no one else would. He accepted me for all my faults and pettiness. He saw the good in me, dear cousin, even when I couldn't find it in myself.

We are moving to California in a few weeks. Michael wants to join his brother in a new business venture. It would be so good to see you before I left. I could come to Rowan, or even Chicago, and we could have lunch or dinner. I am not sure when we will have the chance again. I don't think I will be back.

We used to be close. I would like us to be close again. You were always like a sister to me. We are the only family we have left. I look forward to hearing from you.

From the heart,
Margaret

Lisbeth never opened the letter so she never knew Margaret's side of the story. Hollis frowned. Lisbeth had passed judgment on her cousin, as had she. Hollis felt chagrined; she didn't like finding that out about herself. After all, she had been falsely accused only a few years ago.

Still, Margaret had held on to Lisbeth's returned letter of rebuff. Curious ....

She picked up Pierson's report and turned to the biographical page, which indicated that Margaret had died of natural causes at eighty-eight.

The next letter, dated 1954, had been opened and stamped in the upper left corner with a government seal.

Dear Mr. and Mrs. Koch,

I want to thank you for your generous donation to Congressman Guber's re-election campaign. The Congressman is always glad to serve constituents who are as enthusiastic as he is about this wonderful country. Also, thank you for offering to sponsor a fundraiser in your lovely home.

Congressman Guber received your note of gratitude. He wanted to let you know that if he was somehow helpful in obtaining the release of Eric Ferris, then he was glad to assist. He knew once the governor was aware of Mr. Ferris' circumstances, he would agree that a commutation was appropriate.

We look forward to seeing you in October. If we may be of further assistance, please do not hesitate to contact us.

Sincerely,
Kevin Werthy
Congressional Aide,
10th District

Hollis read the letter three times. Margaret had used her resources to get Eric out of prison. She wondered how her husband felt about that. Or maybe it was the price he paid to have the beautiful Margaret on his arm.

Picking up the last letter, dated 1957, a feeling of finality came over her. It had been stamped RETURNED. She opened it with resignation.

Dear Eric,

I don't have the words to tell you how sorry I am because there are no words. I wasn't the woman for you. I wasn't the woman for anyone. I couldn't bear seeing you in prison. I wanted to remember you the way you were before Charles died. I did you a favor getting the divorce.

I understand you were able to get a release. Hopefully, you went back to living a good life. Hopefully you didn't let our relationship keep you from finding true happiness. Now it's water under the bridge. The events that pushed us apart carried us away in a current of consequences that we never sought or deserved.

I know you're wondering why after all this time I've gotten in touch with you. I've lived a full life, and although I never had any children, I have found peace. But I can't find complete peace until I express my guilt and sorrow at the way things turned out between us. I'm not sure if this letter will ever find you, but I had to write the words.

I am so sorry Eric, for causing you any pain.

I understand you went back to Rowan and then moved to California. I hope you found peace as well.

Respectfully,
Margaret

Hollis took a sip of tea. She gently put the page back into its envelope. Well, George would be glad to hear that their client didn't have any children. Margaret had written to Eric, but his heavy "return to sender" scrawl—and she had no doubt it was his—across the front of the envelope indicated he had had no interest in reading what she had to say. Margaret knew that Eric Ferris had returned to Rowan but chose to settle in California.

Near Margaret?

She had come to the end of Margaret's story. There were no more letters. She wrapped the bundle together with the silver ribbon and headed down the hall.

In his office, George was deep in a volume of appellate cases.

"What's the matter with you?" he said, putting the book down. "Why the glum look? You look like you lost your first case."

Hollis shrugged and sat. "I'm finished with the letters."

"Well?"

"It looks like there are no heirs. Margaret only had one cousin and she passed away many years before Margaret. She was her only family."

George slapped his hand on the desk. "Great. Let's finish filing the paperwork and get a court hearing date."

"Don't forget, I'm meeting with Kelly Schaefer next Tuesday. She says she has another letter."

George shook his head. "Considering the timing of these letters, it's unlikely that Margaret could have squeezed a child in." He noticed the muted response from Hollis. "What?"

Hollis avoided his eyes.

"I've read the letters, and I guess I had a picture in my mind about who Margaret was. She was self-centered, vain, insensitive, and shallow. I didn't think I liked her very much."

"Hollis, you don't have to like—"

"I know that." She ran her hand through her hair. "Anyway, I fell into making assumptions. But the letters got to me. It's different once you know a person's background. Life is so ... fragile."

"I don't know what surprises me more, the fact that your caring has caught you off guard, or that you let someone get under your skin, even though that person is dead."

"George." She took a large swallow. "You make me sound … sound cold and indifferent."

He shook his head. "Not at all. I think you just keep things tightly wrapped inside. Look, you're probably just tired. I know I am. Go home and have a good weekend. We'll aim for Wednesday to file the court documents."

Hollis straightened in her chair and brushed imaginary lint off her sweater. "No, I'm okay. I can have the forms ready to go in an hour. If you look over my draft order, I can work over the weekend to finalize it."

He bent down and gathered his briefcase and sunglasses. "No need to hurry. I'm going home. I have a life to get back to."

Hollis flinched.

George steered her into the hallway. "Go home, Hollis, have some fun. I'll see you on Monday."

She gave him her fake peppy smile.

"I know, I'll practice getting a life."

AT HOME, HOLLIS thought back to the conversation with George. She had to admit that her life was becoming a bit one-dimensional. But things were looking up. She had a date to get ready for on Saturday and a chance to break out of her protective shell. As she had for the past three days, she replayed the conversation with Brad in her mind—what she should have said, what she could have said, what she shouldn't have said. By the time she replayed and replayed their conversation in her head, she was a mental wreck—a mental wreck with nothing to wear on her first date in six months. She'd have to improvise.

"Stephanie, do you think my black sheath is too formal for a Yo-Yo Ma concert?" Hollis asked, holding the phone to her ear and riffling through her closet with the other hand.

Whatever Stephanie was chewing sounded delicious. "I don't think I've ever seen your black sheath."

"You know, the one I wore to your Christmas party. You said I looked nice."

"I was being polite; it was Christmas." Stephanie stopped chewing. "Besides, you wouldn't wear a dress to the Pavilion. You're going to be sitting on grass."

Hollis smiled to herself. "No, we're going to be having dinner there, so we'll have seats."

"Hmm, nice, but it doesn't matter, you still need to wear pants and a sexy top."

Hollis was silent.

Stephanie started chewing again. "I get it. Even if you knew what a sexy top looked like, you don't own one." She wiped her mouth. "Don't worry. You were good enough to help me with my evidence report last week. I've got to drop something off at the post office this evening. I can bring by a couple of potential candidates."

"That's okay; I can be at your house in a short while."

It was Stephanie's turn to be silent.

Then, "Why am I not surprised?"

Hollis ignored her. Somewhere in the back of her mind, she knew her aversion to having people come to her home bordered on neurosis. She had taken a few psychology classes and knew it related to her trust issues—as in lack of. At any rate, Stephanie might be right for the wrong reasons.

BRAD ARRIVED EXACTLY on time. By then Hollis was very pleased with her appearance. She and Stephanie had settled on an electric blue peasant blouse with colorful embroidery along an open V-neck—sexy without being obvious. With her black slacks and matching espadrilles, she felt confident he wouldn't be disappointed.

And judging by the appreciative look he gave her, he wasn't.

"Why, Miss Morgan, you look fantastic tonight." He opened the car door as she came down the walkway. "Are you ready for a great concert?"

"More than ready. I've been studying for the bar and haven't been out just for a good time in months." She smiled. "Thanks for inviting me."

He drove well—not too fast and with confidence.

"Have you known George long?" Hollis asked, to fill in the silence that had fallen between them.

Brad kept his eyes on the road. "I worked an assignment with a mutual friend. When George needed help with the Koch case, he contacted me." He turned and smiled at her. "It looks like it may have been my lucky day."

Their seats were stage front, about ten rows up a slight slope. Dinner was not the best she'd ever tasted, but an evening under the stars made it delicious.

Their conversation lagged until Yo-Yo Ma's breathtaking performance silenced them, and there was no need for conversation at all. During intermission Hollis was ready with her prepared topics.

"So, I've been reading up on Margaret Koch. It wasn't in your file, but I discovered that Margaret was married three times." She didn't want to sound smug, but she still heard it in her voice.

Brad stiffened.

"Really? Did you discover any heirs?"

From his tone, Hollis knew she was on thin ice. "No, you had that covered. I was reading old letters that she left behind—"

"My report stands. She didn't have any heirs. My assignment wasn't to chase out gossip—"

"Excuse me, but I wasn't reading for titillation." Hollis caught herself and took a breath. "Let's start over. Isn't Yo-Yo Ma fantastic?"

She felt his body shift. He laughed.

"Yes, he is. Are you enjoying yourself?" He turned and picked up her hand. "I didn't mean to fly off like that."

Hollis shook her head. "No, I'm the one who brought up work. And, yes, I'm enjoying myself very much." She squeezed his hand then slipped hers free.

She racked her brain to come up with appropriate small talk. She had used up her limited repertoire, and the only topic left was the weather.

"You're a beautiful woman."

"What?" She came back from her own thoughts.

He laughed again. "Tell me, what would you rather be doing right now?"

Oh, oh, a trick question.

"Ah, you answer first."

"I'm right where I want to be, with the woman I want to be with." His blue eyes sought hers.

Hollis fought down panic. "I'm doing what I want to be doing, too." She smiled, and then her smile faded. "Brad ...."

He raised his hand. "No pressure. Let's just enjoy Yo-Yo Ma."

"I agree." She bit her lip. "Would it be really tacky to ask you a question about Margaret Koch, now?"

"Yes, it would." He turned to the stage and shook his head. "But how about on the drive home?"

Hollis grinned and ran an invisible zipper across her lips.

The lights brightened and then faded to dim.

Intermission over, Yo-Yo Ma took the stage again. Hollis let herself sink into the mellow notes of the cello master. Her eyes closed, but from time to time she could still feel Brad's gaze on her.

Afterward, they walked to the car at a leisurely pace, commenting on the memorable performance. Surprisingly they were out of the lot in record time.

"Want to go for a drink at Crowley's?" Brad offered.

Crowley's was a neighborhood bar and restaurant in Walnut Creek. Its view of the hills and themed Italian décor made for a classy, quiet and friendly atmosphere. It was Hollis' ex-husband's favorite hangout, but she didn't hold it against the owner. She and Brad settled into a booth toward the back. Hollis ordered Pinot Grigio, and Brad a screwdriver. The drinks came quickly, and after making sure they wanted nothing else, the waitress left them alone.

"All right, you've kept your word. What do you want to know about Margaret Koch?"

Hollis had been fiddling with her cocktail napkin, waiting for him to bring up the subject. "I've been reading these letters sent to Margaret. They leave an image of a cold, selfish but attractive woman—but also someone with loyal friends who helped to get her ex-husband out of prison." Hollis rubbed her forehead with her fingers. "What's, or what was, your impression of her?"

"Well, obviously I never met the lady. She'd been married to Koch for over forty years, so I didn't know about her ex-husbands. I pored over the public records. There were no living relatives and no heirs." He waved the waitress over and pointed to his glass. She acknowledged him with a little wave and headed for the bar. "You're right; I didn't follow up on the first two marriages. But if she had any children I would have picked it up under Koch."

"Brad, I'm not talking about what's in your report, I'm talking about what did you think of—*her*?"

He cocked his head.

"She was a beautiful woman—smart, too. Her signature was on all the business documents and income tax forms. I got the sense Koch catered to her. He was nineteen years older." His drink came. "I don't know, I guess I would say she'd be a catch."

Hollis gave him a quick smile in understanding.

"A catch, huh." She played with her napkin. She noticed he was looking around the restaurant. "Tell me, Mr. Pierson, what do you do in your spare time? Any hobbies?"

"Football. Can't live without it."

Hollis grinned. "Do you play?"

He shook his head and laughed. "Nah, I just watch."

"And when football season is over?"

"I go over game tapes on my TIVO."

Of course.

She leaned forward. "Tell me, how did a football junkie like you ever hear about Yo-Yo Ma?"

Now he looked sheepish.

"My mother has season tickets. She suggested the concert."

"Oh."

"Hey, what about you? What do you like to do in your spare time?"

Hollis shrugged. "I read. I love good music. I love good wine."

"Hey, I love wine too."

Hollis looked pointedly at his glass.

Brad smiled. "Well, not all the time, but I enjoy a good cabernet. Maybe we can go to the wine country sometime?"

Hollis' eyes locked with his. She felt a shiver. "I'd like that."

He touched her fingers lightly to his. "Do you think we could do it before pre-season football?"

The glint in his eyes didn't escape Hollis' notice. She laughed, throwing the cardboard coaster at his head.

CHAPTER THIRTEEN

H OLLIS COULD TELL by Detective Cavanaugh's furtive glances at his desk clock that he had mentally moved on to his next meeting. She reached across and turned it to face outward.

"Ms. Morgan I have other app—"

"I'm sorry if I'm holding you up, but I'm telling you: Cathy was murdered for her article. Everything points to someone who wanted her dead. Someone who doesn't want the Fields' story to come out." Hollis sat on the edge of the chair and pounded the top of the desk with her fist.

"Ms. Mor—"

"I'm not just ranting. Did you speak with her assistant at *Transformation* magazine? Did she tell you that Cathy was threatened?" Hollis knew she shouldn't raise her voice, but she couldn't stop herself. "A couple of days ago someone broke into my condo. Nothing of obvious value was stolen. I don't know what he was looking for, or even if he found it, but I'm telling you, there should be some very high profile suspects in this case." Spent, she sat back in the chair.

"If you would just let me speak, Ms. Morgan." He turned the

clock back around. "Detective Faber told me of your break-in. Unfortunately we can't find a connection be—"

"Wait. Look at this." She put a small plastic bag with the notepaper she had found in her condo. "See? They knew I lived there. I was targeted."

Cavanaugh picked up the bag and peered at the paper. He frowned and put the plastic bag and its contents in his side drawer.

"You said Ms. Briscoe had a personal assistant. How did you find her? No one at *Transformation* mentioned an assistant. What's her name?"

Hollis licked her lips. She wasn't ready to acknowledge she had kept copies of the notes that Cathy left with her, but Mark's words of warning came back to her.

"She wasn't a personal assistant. I meant to say she assisted Cathy." She rushed, "Look, I made a copy of Cathy's notes. You have the originals. I only discovered her through a fluke—by running into her in the bathroom. *Transformation* management didn't tell us, either." She paused. "Her name is Gail Baylor."

He wrote down the name.

Hollis feared her growing impatience would turn Cavanaugh against her. "Will you please consider that maybe Cathy had something that a celebrity like Fields might want?"

His blue eyes seemed to look through her. She could tell he didn't believe her.

"Tell me, Ms. Morgan, you're a student of the law. Are you aware of California Penal Code Section Thirty-two?"

She could feel a rush of blood climbing up her chest.

"I'm not withholding information to obstruct justice," Hollis said. "I want justice. I want to make sure you look in all directions regarding Cathy's death."

"And so you—"

"All right," she sighed. "And so I made copies of Cathy's notes. On one page there was a phone number that led me to her assistant."

He leaned back, a satisfied smirk on his face.

"Does Miss Baylor consider the threat to be real?"

Hollis hesitated. The last thing she wanted was to make trouble for Gail. Withholding information could be just as problematic as being an accessory after the fact.

"No, I don't think so. She's the real nervous type. I don't think she linked the two."

Cavanaugh continued to stare at her. "But you did?"

"Detective, I think it's obvious. We never found Cathy's research materials. Any writer, let alone an attorney, would have her research saved to back up any claims. Her final article was really more of a tease for a future series on non-profits. I think Cathy was murdered by someone on Dorian Fields' staff. I doubt that Fields would do it himself. That person killed Cathy and took all her incriminating research material. You asked me; now that's what I think."

It was her turn to lean back in her chair and cross her arms.

He flipped his pencil over and over, point to eraser, eraser to point.

"Is there anything else you're withholding?"

"There is no 'else.' I never withheld anything. I'm on a legal team representing *Transformation* magazine. We have a right to collect information in defense of our client." Gathering her purse and folders, Hollis rose to her feet. "I didn't have to come here."

Cavanaugh walked around his desk and blocked her exit. "That's right, you didn't. But if you really think you have pertinent information about your friend's murder, then you will serve her best by working with us."

She gave him a broad smile. "Then you'll look into Dorian Fields' role in all this?"

He lightly touched her elbow, guiding her to the door. "Remember what I said. Work with me on this. We already spoke with Miss Briscoe's assistant, but we'll go back and ask her about the information you just gave me."

Hollis nodded. She was worried that Gail would not like being questioned about their conversation. She remembered the call Gail took. Hollis knew it had something to do with Fields, but she wanted to make sure her hunch was correct. She had no illusions about the lightly disguised threat.

She allowed herself to be ushered to the other side of the door, which Cavanaugh pushed closed behind her.

HOLLIS COULDN'T GET back to her office fast enough, but after fifteen minutes on the phone she was still trying to convince Gail of her intentions.

"I'm not questioning your motives; I just want to know if you made a copy of Cathy's research." Hollis bit her lip to keep her exasperation in check.

"Cathy trusted me, and I wanted to make sure I didn't overlook any text. Resorting to memory isn't professional."

"Gail—"

"I never knew when or where she might call me."

"Gail—"

"Offices get hacked all the time and I didn't want to be caught—"

"Gail, please."

"Yes, I have a copy."

Hollis took a deep breath. "Good, I'm going to need that copy. Can you email it to me?"

She looked up to see George looming over her desk. *How long had he been there?* She turned slightly away from his gaze and put the phone to her other ear. He took the hint and left.

"Are you still there?" Hollis said.

"I have to think about this. Cathy never wanted to save her work on our system. She was afraid of it being stolen. I don't have it on our server."

"Well, then, where is her research?"

"It's on a thumb drive."

"Okay, I'll come by to pick it up."

"No, I don't want management to think I'm disloyal. If you come by again, they will suspect we are working together." Gail paused. "I'll mail it to you."

Hollis was glad Gail couldn't see her frown of disbelief. "Gail, we are working together. I'm defending one of your writers." She mentally counted to six. "Look, I don't trust the mail. Let's just meet somewhere."

There was muffled dialog. Someone had evidently entered Gail's office.

"Oh, my God, the police are here to question me." Panic was clearly evident in her voice. "I don't want to be arrested. I've got to go."

Hollis stood. "Wait! Mailing the thumb drive is fine. The police may want a copy too, but make sure you put one in the mail to me. Okay?"

"What? Yes, no problem. Goodbye."

"Wait, you don't have my address. It's—"

But Gail had already put down the phone.

HOLLIS ARRIVED AT Mark's office late, but he was too preoccupied to notice. They went through copies of her notes and Mark pored over Hollis' draft court motions.

She pulled her sweater close. The air conditioner seemed to be set to just above freezing. Maybe it was just to impress clients, because there was little need for air conditioning in the Bay Area.

Hollis recounted for Mark her phone call with Gail.

Mark raised his head. "That thumb drive is critical. It could be the answer to all our questions."

She nodded. "If what Gail told me is true, having Cathy's research takes us from a possible lose to a probable win," Hollis replied. "We have a lot going for us. I just know those non-profits are bogus. Wherever Fields' money is going, it's not going to them."

"I agree that's what it appears to be, but we need to get the court to see it our way." Mark stuffed pages into file folders. "Let me know as soon as you hear."

"If she puts the thumb drive in the mail today, I should have it by day after tomorrow or maybe even tomorrow. I'll let you know as soon as I get it."

"I read through the questions you drafted for the depositions; they're excellent. You're going to make a great attorney, Hollis."

"Thanks, I hope the State of California agrees with you." Hollis couldn't hold back the smile that eased across her face. "Are you sure you don't want me to assist? I've met all of them and it helps to have a second pair of eyes."

Mark shook his head. "Nope, you go ahead and wrap up the Koch case. You'll be able to get George off your back. Besides, I would rather have your help with the settlement hearing."

"Don't worry about George, he's happy. Margaret's letters indicate she doesn't have any heirs and we can close the matter."

"So why bother with Kelly?"

Hollis glanced at Mark and then looked away. "She has one more letter, and I'm trying to get her to let me read it."

"Again, why?"

"Because … because Margaret's story has me hooked. I'm not being nosy. I just want to … I'm hoping … to find out what happened."

How could she explain to him what she couldn't explain to herself? Just then his desk phone rang and she took a deep breath. From his end of the conversation, she could tell it was the front desk.

Mark laid down the phone. "My client is here. I've got to go." He gathered up a legal pad and pen. "When you get Cathy's thumb drive, start transcribing and …."

He saw the look of irritation on her face and stopped.

"Mark, if you don't want me to leave you working this case alone, don't even pretend to tell me how to do my job."

"It's the stress." He pretended to duck from her imaginary blow.

***

HOLLIS USED THE time on the trip back from Mark's office to consider her next move if she didn't hear from Gail. When she got back to the firm she would call every fifteen minutes if necessary. She had to get her hands on that thumb drive.

She rushed through the lobby.

"Hollis, wait!" Tiffany called to her. "You had a visitor. She wouldn't leave her name. She said she couldn't stay but would call you in the morning."

Hollis' heart took on a rapid staccato. "What did she look like?"

"Older, kind of scattered and really nervous."

Gail.

"If for any reason she calls or comes by again, and I'm not at my desk, can you come and find me? It's truly important that I speak with her."

"Sure." Tiffany paused. "I think she had a package for you, but she held on to it. I gave her your card."

Hollis voice sounded strained to her ears. "Was it a small package?"

Tiffany nodded slowly.

Hollis started for her office again. She didn't have a good feeling about missing Gail. The woman might take it as some kind of omen that Hollis hadn't been available. She picked up the phone and dialed Gail's number, but there was no answer.

By four-thirty Hollis wanted to scream into her purse.

"Tiffany, I have to go to the mail room. I'll bring the office mail back."

It was a long shot, but she was in desperate need of a distraction.

"Thanks." Tiffany's look of surprise did not go unnoticed.

Hollis hated the mail room. It was a maze of right and left turns that usually brought her back to where she started. At least this time she could hear machine sounds and she followed them carefully. The hallway opened into a large windowless room.

A fan whirred loudly and she had to raise her voice over the noise.

"Mail for Triple D?"

The young man with an iPod bud in one ear nodded, pointing her to a long table covered with a dozen white square plastic containers brimming with mail. She quickly went through the firm's half-filled box and sighed.

Gail's package was not there.

HOLLIS TOSSED AND turned through the night, berating herself repeatedly for not getting that thumb drive. She should have been in the office when Gail came by.

The next morning she went to work planning to go straight to *Transformation's* offices, but remembering how Gail had been so uncomfortable being seen with her, she decided to call from a nearby coffee shop and ask her to come down. They needed Cathy's research, or else Fields' attorneys would tear their case apart.

Even though she wasn't sure how things could have happened differently, without that thumb drive she felt she had let Mark, and more importantly, Cathy, down. She checked her voicemail for the third time and then picked up her tote to leave for lunch.

Her phone buzzed, startling her.

"Yes, Tiffany?"

"There are two detectives here to see you, Hollis. Shall I put them in the conference room?" Her voice sounded shaky.

Hollis groaned.

What now?

"Yes, that'll be fine, put them in the small conference room. I'll be right out."

She could have told Cavanaugh on the phone that Gail still hadn't given her the thumb drive and spared him the trip to her office.

She couldn't hide her expression of surprise when she

entered the conference room.

Cavanaugh and Faber stood as she walked in. "Ms. Morgan, thank you for seeing us. I brought Detective Faber because it seems he has an interest in this new case as well." Cavanaugh took a chair.

Faber nodded her way. She thought she caught a smile, but it was gone too quickly. She sat across from the detectives.

"What new case?" she said.

Cavanaugh pulled out a small notebook. "Yesterday evening, along Skyline Ridge, a body was discovered by a park ranger. We have since identified the body as Gail Baylor."

"Oh, no," Hollis gasped, and put the back of her hand to her mouth. "How was she killed?"

"We're still checking into that," Cavanaugh replied. "But tell me, the other day you seemed convinced Ms. Baylor had information we would find useful in the murder of your friend. We want to know exactly what you hoped to find."

She frowned and glanced over to Faber, but his face wasn't revealing any clues.

"The first thing you need to know is that Gail came to see me yesterday afternoon."

The detectives exchanged looks. Faber slipped out his notebook.

"Ms. Morgan, I warned you—"

Raising her hand, she stopped Cavanaugh's sentence.

"Wait! Before you get all wound up, I didn't know she was coming here. I never met with her. I missed seeing her. I'm not even positive what she wanted." Hollis clasped her hands to keep them from shaking.

"What time was she here?" Faber spoke this time.

Hollis shrugged. "I had a one o'clock meeting out of the office. I was back by two o'clock."

Faber asked. "I understand you had just met with Ms. Baylor?"

"That's right," Hollis said. "Cathy never mentioned her. I

stumbled on her ... I think it was Cathy's notes." At the last moment she remembered her earlier statement to Cavanaugh.

"Why didn't she come forward when she heard about Briscoe's death?" Faber asked.

She rubbed her forehead. "I don't know. Gail worked for several of *Transformation's* writers. Cathy Briscoe was just one of her assignments."

Hollis hoped the frustration in her voice wasn't too apparent. It didn't seem to matter, though. Cavanaugh didn't even glance her way.

"I see," Faber asked. "What do you think she wanted to see you about?"

She looked away. She wasn't ready to give up her theory about what she found in Cathy's file; it was still just a theory. But she also remembered Cavanaugh's warning.

"There is a thumb drive that might have Cathy Briscoe's research on it. Mark Haddan and I think it will prove Cathy's contention that Dorian Fields is a crook, and it should also give a probable motive for Cathy's—and I guess now, Gail's—murder."

She had their attention now.

Cavanaugh began to pace as he talked. "We collected quite a bit of material from Ms. Baylor. We are still going through it all. However, there were a number of thumb drives. It seems she would keep backup documents on each subject file. We did not find one for Catherine Briscoe. Where's Miss Briscoe's thumb drive?"

Hollis leaned across the table. "I don't know. I think Gail was bringing it to me yesterday. Our receptionist said she carried a small package, but she wouldn't leave it."

Cavanaugh's face looked grim. "Did you encourage her to withhold information from the police?"

"Of course not."

"Were you expecting her to come to your office?"

She shook her head. "No, the only thing I can think of is that she must have gotten my address from our first meeting, when I told her I worked for Triple D. We had only talked about her mailing the drive, not dropping it off."

Cavanaugh stopped pacing. He sat down and started to tap the top of the table with his pen. "I take it you've checked the mail?"

"Yes."

Cavanaugh placed his palms on the table and leaned toward her. "I know you and Haddan took up your friend's court case, but if you come across any information relative to either victim—and I mean *anything*—I want you to contact me immediately. This is now a double murder case."

"Don't worry, Detective. You'll be the first to know."

Gail's murderer had the thumb drive, and the three of them knew it.

CHAPTER FOURTEEN

~

FROM HER DESK drawer, Hollis pulled Margaret's returned letter to Eric Ferris. She needed to finish the Koch matter so she could concentrate on Cathy's murder. There were just a few loose ends she wanted to clear up before her meeting with Kelly at the end of the week.

Ferris' address was in Vacaville. Using a reverse phone number search online, it took a few calls to back trace the address to a phone number. A young woman who had been renting there for less than a year answered. While she knew Ferris was her landlord, she and her husband communicated with him through a property management company.

"Can you give me the name of the property management company?" Hollis asked after introducing herself.

She heard the hesitation as the young woman tried to decide if this was a serious call. "My husband handles all the family business. I really can't help you."

"I'm sorry, I didn't tell you. The law firm I work for, Dodson, Dodson and Doyle in Oakland, represents the estate of a client who had a connection to the former occupant of your home. We're just trying to locate him."

Hollis sensed the woman's comfort level increase now that she had information she could verify.

"If you leave your name and your firm's office number, I'll have my husband get back in touch with you."

Hollis gave her the information.

"And your husband's name is ….?"

"Alvin. Alvin Gregory."

"Well, thank you, Mrs. Gregory. I look forward to hearing from your husband."

She could hear a sigh.

"Oh, forget all that. Alvin's got a big project at work and he'd only tell me to call you back anyway. I'll leave the number of the property management firm for you this evening on your voicemail. I've got to find it first. You'll have it when you get in tomorrow morning."

Static came on the line. Mrs. Gregory must have been walking around with the phone. "I can tell you this, though. The house was a mess when we moved here. We've been fixing it up 'cause we'd like to buy it someday. If it's part of an estate, I can tell you we increased the value a ton. Alvin had an appraisal done before we moved in and—"

"Mrs. Gregory, your house is not considered part of the estate. We think the prior owner may have an interest in our client's estate, and we're just trying to locate him."

"Oh, oh, okay. Well, we were told they put him in a home in Fairfield. He didn't have any family here. He must have lived by himself. Like I said, this place was a mess. You could tell he didn't have a woman in the house, or at least not for some time."

THE NEXT MORNING Hollis double-checked the lock on her front door, tapped the address of the retirement home into her car's GPS, and headed for the freeway. With luck she would pull into Fairfield before noon. She left a message for George telling him of her field trip, but only when she was positive he wouldn't answer in person.

In fact, the ride was uneventful and really quite pretty. She felt her shoulders relax as the trees got more plentiful and the commercial centers more spread out. It wasn't long before she pulled into the parking lot of the Eastbrook Hills Residence Park. She noted that it was not on a hill and not in a park. The unimpressive three-story building rose solidly from a flat parcel at the rear of what looked like a mid-size shopping center. Two palm trees stood tall on either side of the front walkway. They were the only palms in the area and stood defiantly among the native oaks and fir trees.

Only three other cars were in the parking lot. She walked purposely up to the front entry. The lobby was clean, with a slight medicinal smell. Classical music played in the background. Sun beamed through large multi-paned windows. There was no one in the lobby area.

She lightly tapped the gold Asian-style bell on the reception counter with a name plate that read, Miriam Coulter. A low murmur of conversation from an inner office suddenly stopped, and an older woman emerged. Her obviously dyed black hair settled like a helmet on her small head.

Smiling broadly at Hollis, she said, "May I help you?"

"Yes, thank you. My name is Hollis Morgan. I would like to visit with Eric Ferris. Is he available?"

Her plastered smile didn't break a crack. "So, is he expecting you?"

"No, I was in the area." Hollis gave her a wide-eyed look. "I hoped I could just drop in." She deliberately hadn't made an appointment. She wanted to catch him off guard.

"Are you a relative?"

"No, I'm a friend of an old friend."

Sort of.

"I see." She rolled her secretary's chair over to a Rolodex on a side table and flipped the cards until she came to the one she wanted. Beaming Hollis another of her Stepford smiles, she asked, "Do you have a business card?"

Hollis reached over the counter and handed her a card.

Without looking down at it, the woman said, "I'll have to see if Mr. Ferris is available to meet with you. Please have a seat."

She pointed Hollis to a row of stiff-backed oak chairs lining a short dividing wall.

Hollis glanced at the stack of *Senior Living* and *Retired Life* magazines on a small round table, but before she could read about cruises for seniors, Miriam was back.

"I'm sorry, but Mr. Ferris wasn't expecting you and is unable to meet with you right now." She did seem sincerely sorry. "However, he asks that if you are amenable, he can meet with you after lunch, say one-thirty?"

Hollis was pleasantly surprised he was willing to meet with a stranger, but then she couldn't imagine he would have many visitors.

She stood and put her purse under her arm. "No problem. I'll run a couple of errands and return then."

Fairfield appeared to be a quiet suburban community with its growth emphasis on retail. Hollis chose to have lunch in a simple looking café about five minutes from the retirement home.

She checked her voicemails and then left her own message.

"Mark, I'm ready to go ahead with the Martini deposition tomorrow. I'll be there early and meet you in your offices. I'm not at work today so you won't be able to reach me." Hollis hit the send button. Hopefully Mark would get her message before he went into the meeting with *Transformation*.

He returned her call within seconds.

"Where are you?"

Hollis thought about lying but she was trying to practice telling the truth. "I'm in Fairfield. I'm tracing down Margaret Koch's ex-husband."

"Why?"

"I'm just making sure there are no heirs." She paused. "No, forget that, the truth is I just want to see what kind of ex she had. I want to put faces to the people in the letters."

She could hear Mark riffling through papers.

"Not to knock your detective instincts, but I could really use your attention on Cathy's case. I want to be ready to depose Ms. Martini, and I could use your help with some additional questions."

His tone irked her—not because of what he said, but because he was right.

She added, "I'll have a supplemental set of questions to you by five .... No, make that by four o'clock this afternoon."

RETURNING TO EASTBROOK Hills, Hollis found Miriam sitting primly at her desk, waiting for her to return.

She smiled. "Hello again, Miss Morgan. Mr. Ferris will meet with you in the library. I can show you the way."

Hollis murmured her thanks and followed the woman down the carpeted hallway. Eastbrook also had a senior day center that looked much nicer than the one in San Lucian where she volunteered free legal assistance. Impressionist reproductions lined the walls leading to the doorway of the library, which appeared to be well stocked with books. A circular walnut table occupied the center of the room, while overstuffed chairs were situated in small alcoves that offered privacy.

Eric Ferris sat in one of the alcoves farthest from the door. A thin, tall man with strands of light brown hair crossing his scalp, he was stooped over a book, but Hollis got the feeling he was following her with his peripheral vision.

Miriam lightly touched his shoulder. "Mr. Ferris, this is Miss Morgan. I'll leave you two," she said, then to Hollis, "Miss Morgan, if you could let me know when you leave, I'll be at my desk."

He looked up, his expression blank, and placed his book down on a side table.

"So, to what do I owe this pleasure?" His voice was deep and so melodic it was disconcerting. Had she not been standing in front of him, she would have sworn it was the voice of a much younger man.

"Mr. Ferris—"

"Eric."

"Okay, Eric, my name is Hollis Morgan and I work for a law firm that is handling the estate of Margaret Koch." She licked her lips. "I wanted to—"

"Who did you say?" His voice no longer held any warmth.

"Margaret—"

He closed his eyes and leaned his head against the chair. His hand trembled as he rubbed his chin. Hollis said nothing. After a moment he turned his pale blue eyes to her.

"What do you want?" His tone had an edge of danger and anger. Neither of which Hollis wanted to pursue.

Suddenly, she felt foolish and intrusive. She should not have come. This was just plain nosiness. George would reprimand her if he knew what she was doing.

"I just wanted to give you this letter—"

"Get out. Leave me alone." He waved her away.

Hollis held out the letter. "But here I—"

He rolled his chair around, bumping into her leg and knocking the letter to the floor. The wheels of the chair made marks across the envelope.

He called out, "Nurse! I want to go to my room, now."

HOLLIS PATTED A large client file of briefs, orders, and motions. It was the final filings for Margaret Koch. Before the day was over she would give the bundle to George for his review and signature. After yesterday's failure with Eric Ferris she was more than ready to close the matter. Hollis gathered up her pen and pad. She would enter the paperwork with the Probate Court tomorrow. The hearing would be in thirty days.

She glanced at the clock. She had thought it was important to have the visit to Eric Ferris accomplished before she met with his granddaughter, but now she wasn't so sure it made a difference.

Hollis told Tiffany that she would be in her office, and when she arrived, to take Kelly Schaefer directly to the conference room.

About fifteen minutes later, her phone buzzed. Her guest was waiting. She scooped the box of letters under one arm.

Looking forlorn, Kelly stared out the window. As Hollis reached out to shake her hand, she had a good look at the young woman. The *Town & Country* style Hollis noticed at their first meeting was still present. Kelly wore her sandy brown hair long and off her face. The pair of glasses made her look older; Hollis wondered if they were necessary or merely a prop like the ones so many young people wore these days.

"Ms. Schaefer, we meet again." Hollis directed her to a seat at the table.

"You found me, and you found out my phone number. What else do you know about me?" Kelly said.

"I know you're related to Eric Ferris. My guess is that you're his granddaughter."

"So you know about him, too." Kelly took a deep breath. "Yes, I'm his granddaughter."

Hollis wasn't really surprised. The picture in the puzzle was starting to fill in.

"Will you tell me a little about yourself?"

Kelly's brow wrinkled, but she acquiesced. "There's not that much. I grew up in Chicago. My mother was a Ferris. She died when I was eight from renal failure. I was raised by my dad until he died when I was sixteen. Then it was just me and Grandfather."

"Do you still see him?"

Kelly nodded.

"Why were you looking for Margaret Koch's letters?"

"I didn't know what I was looking for," Kelly said. "Something to prove my Grandfather was innocent and shouldn't have gone to prison for a crime that Margaret Koch committed."

Hollis sat a little straighter. "Wait a minute. There was no evidence that Margaret killed Charles Ferris."

"Oh, yes she did. My grandfather would never talk about it, but my dad told me what happened." Kelly's voice quavered. "My Grandfather saw her kill Charles and then he sacrificed himself and went to prison for her crime."

Hollis said nothing. The tension in the room was palpable. Hollis finally spoke when she saw Kelly's shoulders drop in resignation.

"You said you had a letter. Can I see it?" Hollis asked. "Here are copies of mine."

Kelly pulled her purse closer to her. "I found this letter among my mother's possessions when I was packing Dad's things. It has a lot of sentimental value." She motioned with her head. "Now my grandfather is in a care facility, and I know he doesn't want to die with a crime attached to his name. It's what got me started looking in Margaret Koch's house for proof that she killed Charles Ferris." She stared at Hollis' box of letters. "I've never let anyone else see it."

"I think I understand. Before we go there, there's a lot about Margaret Koch as a person that confused me. But like most people she wasn't all bad, or all good." Hollis put her hand on the box of letters. "Er ... I went to visit your grandfather. I didn't know about your relationship."

"You saw Granddad? How was he?"

Hollis flinched. "Well, I think I may have upset him."

"You upset Granddad? What did you say?" Kelly's concern was evident.

"I tried to give him a letter from Margaret that she sent to him years ago, but he refused to take it."

Kelly shook her head. "Is there some reason why you can't mind your own business?"

Hollis could feel her defenses starting to rise, but then she knew that Kelly was right.

She spoke with deliberateness. "I was doing my job. I can tell you now: it's clear from the letters in this box that Margaret thought your grandfather had killed his brother."

Kelly glared and raised her voice. "I don't believe you. Margaret Koch is ... was a liar."

Hollis waited for Kelly to regain her calm.

"To show good faith, I'm going to leave you with the copies of Margaret's letters that relate to your grandfather. There are more but they're not relevant. I would ask that you read them here and have the receptionist come and get me when you're finished. Then maybe you will let me read your letter." Hollis placed the envelopes on the table. "Agreed?"

Kelly nodded. "Agreed."

HOLLIS WENT BACK to her office and tried to distract herself from wanting to be in the conference room looking at Kelly's expression as she read. Instead she prepared a backup set of supplemental questions for Mark's deposition that afternoon.

Her phone flashed a familiar number.

"Mr. Pierson, to what do I owe this call?" She said with a tease in her voice. "It's good to hear from you."

He chuckled. "I can't talk now, but I was wondering if you'd be interested in going on a boating picnic this Saturday?"

"What, isn't this pre-season? Aw, wait, no games on Saturdays, right?"

"Touché." He lowered his voice. "I had a great time at the concert. I'd like to see you again."

Hollis put aside her playful tone. "I'd like to see you again, too. Saturday sounds like fun. I've been told I need to get a life. What time do you want me ready?"

After making the date arrangements, she murmured goodbye.

"Who was that?" Tiffany stood in her doorway.

"Absolutely none of your business." Hollis blushed. She wasn't ready to become the next topic in the office gossip mill.

Tiffany gave her a knowing smile. "I came by to tell you that Miss Schaefer is ready to see you."

Sitting in the conference room and staring into space, Kelly appeared pale and tired. Even though Hollis hadn't given Kelly all the letters to read, she must have sped through them.

"Do you believe in forgiveness?" Kelly asked.

Hollis frowned. "I don't understand your question. What do you mean?"

"My grandfather doesn't believe in forgiveness."

"I don't think belief has anything to do with it. Forgiveness is a gift you give yourself. It allows you to move on." Hollis looked through the window at the glittering bay. "People can do terrible things to each other, and to themselves. Forgiveness is like a reset button."

"A reset button, huh? I like that. But, what I still don't understand is, if Margaret Koch didn't kill Charles and Granddad didn't kill him—who did?"

"I've got an idea about that. That's why I want us to ride out to the retirement home and visit your granddad together. We can ask him what he thinks."

"You're kidding? Just go ask him?"

"Yes. Why should we spend unnecessary time guessing? Technically, my firm is ready to close Margaret Koch's estate. There's one final court appearance." Hollis glanced down at her cellphone, which had been going off repeatedly for the last fifteen minutes. It was George. He was out of the office, but she knew what he wanted. She was avoiding him until she could close the book on the Koch matter. She pushed SAVE. "We could go out there first thing in the morning and get this thing cleared up. You would be my ticket."

Kelly sounded confused, "Ticket?"

Hollis bit her lip. "Well, the last time I visited him, I didn't leave on the best of terms. Besides, I know he would want to see you."

"I can't go this weekend. I've got previous commitments. It will have to be next week."

Hollis didn't know if she could stall George that much longer.

But at least now he could see she had done all the paperwork and the hearing date was scheduled.

"Can we go on Monday?"

"Granddad would know something's up. I usually go on Wednesdays. He likes me to come on Social Wednesday."

"Okay, let's do it Wednesday. We can drive out together." Hollis held out her hand. "Can I see your letter now?"

"I lied. I didn't bring it with me." Kelly frowned. "I know I said okay, but I would rather wait for you to read it after you hear Grandfather's side first. Then I'll share it with you, I promise. I'll drive myself there."

"I see." Hollis mentally kicked herself for not doing a letter exchange. "Okay, so you'll arrange the visit for us. The facility staff know you."

"Well … I guess."

"You contact your granddad. Tell him we'd like to see him. You pick the time and let me know. I'll meet you there."

"What if he doesn't want to see us, well *you*, I mean."

"Oh, I think he will."

CHAPTER FIFTEEN

⟡

A N HOUR LATER, Hollis was sitting quietly next to Mark. Depositions were by nature orchestrated to intimidate. The conference room at Mark's law firm was comfortably furnished and offered a panoramic view of the bay. Unfortunately, the drapes were closed and the room dimmed. Along one side of the table sat Mark, representing the defendant. Along the other side sat Lilia Martini, witness for the plaintiff. Next to her was the attorney for Fields of Giving, and at the head of the table sat the court recorder whose sole job it was to take down every word spoken for the record. These sessions were always tense since one side did everything it could to keep any real information from getting to the other side. They had been there for almost a half hour establishing Lilia Martini's hire date, her function, and the type of clientele being served at her facility.

"Lilia Martini, do you remember meeting Miss Morgan, the woman sitting across from you now?" Mark spoke without looking up from his notes.

Fields' attorney nodded for her to speak.

"Yes, she came to Open Wings the day I got back from

vacation. She seemed nice. I didn't know it would end up with me here."

"Well, do you remember telling her about the activities at Open Wings?"

"Yes, she wanted to know what we did and I told her."

Mark glanced at Hollis, who gave a slight nod.

"How many employees did you tell Ms. Morgan were employed at Open Wings?"

"I told her one, because that's how many there are. Just one, me."

Mark reached into his briefcase and pulled out an annual report. Lilia Martini's eyes grew large. The Fields' attorney appeared to notice a shift in her energy, but it was clear from the confused look on his face that he didn't understand her reaction.

"Ms. Martini, do you recognize this document?"

She nodded slowly.

Mark said, "You need to speak for the recorder, but she will note that you nodded in the affirmative."

"Yes, it's our annual report."

Mark showed the report to the attorney, who acknowledged that it was indeed the annual report.

"Did Ms. Morgan point something out to you in this annual report when she visited?"

Ms. Martini nodded again, then said, "I mean, yes."

Mark flipped through pages and turned the annual report to face her. "Can you read me the section under 'Administration,' marked 'Staff'?"

The attorney reached over and pulled the report toward him. His face flushed, but he said nothing, sliding it over and motioning for Ms. Martini to answer.

"It says that there are five full-time employees who work ... but I tol' Miss Morgan that Mr. Fields' assistant, Miss Phyllis, told me to sign my name. She—"

"My client answered the question. Can we move on?" Her attorney stood.

Mark appeared to look through his notes. "No need, that's it."

The attorney sat back down. "I assume now you're ready to depose Mr. Fields' accountant, Phyllis Meyer?"

Mark returned Hollis' smile. "You got that right."

CHAPTER SIXTEEN

S ATURDAY AFTERNOON WAS mild, and the park's recently mowed grass smelled of coming spring. Hollis took a deep breath and felt the light breeze.

Brad touched her on her cheek and she jumped.

"Hi," she said with a slow grin.

"Hi."

"You were asleep when I came back from putting the boat away."

Hollis nodded. "I've never been in a row boat. Being out on the water felt wonderful. I feel rested."

"Good, I wanted you to have a good time."

"Are you feeling ignored?"

"Not at all. I'm enjoying watching you. This picnic seems to agree with you." He turned over on his stomach. "You know, being outdoors agrees with me too. Other than football games, I haven't done this in years."

Hollis opened her mouth for a retort, then closed it when she saw Brad's grin. "Very funny, actually. I'm proud of you for not mentioning football the whole afternoon."

"I'm proud of you, too, for not mentioning work."

Hollis frowned. "It's been hard. I can't help but wonder who the real Margaret Koch was. The letters she kept were really reminders of her faults and in some cases poor judgment. You'd think she wouldn't want mementos from those days."

"Hmm, from what you've told me—and remember I said we weren't going to talk about work—she may have felt guilty."

She turned to face him. "But you went through her life, looking for heirs. Couldn't you tell that her life—"

"Hollis, I don't mean to cut you off, but I wasn't kidding. I really don't want to talk about work." He rested his chin on the palm of his hand. "Now, what do you want to do next?" He ran his finger along the length of her forearm.

She pulled back. "That tickles."

"Why do I get the feeling we're starting a downward spiral?" He lay on his back.

She drew a deep sigh.

"It's me. I'm sorry." She ran her hand gently over his hair.

"No sorry allowed." He rose up onto his elbow. "Let's change the subject. You know I like football. Tell me, what do you like to do in your spare time?"

Hollis closed her eyes. "It's been so long, but I love to curl up with a good book. I used to belong to a book club and I really enjoyed talking about books, and words, theme and motivation, and the—"

"Books, huh? Figures."

"What do you mean by that?"

"Nothing, but the last book I read was *Old Yeller* in grade school."

"You can remember that far back?"

"Yep, it was the only book I ever finished."

She shook her head. "I don't know about you, Brad."

He reached for her hand and her fingers clasped his.

"They say opposites attract. Maybe we can find some common ground out of the library and off the football field."

Hollis looked down at his ring finger. "Have you ever been married?"

He pulled his hand back and sat up.

"Yeah, and I have the spousal support to prove it." He got a beer from the cooler. His voice had turned frosty as the can. "Let me guess; you only want to be involved in a committed relationship."

Hollis bristled. "Hardly. I've done the divorce wars too. I was only trying to get to know you." She picked up her plate and cup and walked over to the garbage bin.

"I'm sorry," he called after her.

When she returned she stood over him. He put his hand loosely around her ankle.

"I guess this means we won't be going back to my place."

"If it's any consolation, we never were."

He nodded in understanding. "So, what is it be, friends?"

Hollis gave him a patient smile. "Friends."

ON MONDAY MORNING Hollis arrived at her office energized. The turning point with Brad made her realize that she still wanted a solid relationship, but she also noticed it didn't bother her as much as it would have years ago, when her divorce made her question her judgment as well as her attractiveness.

Reluctantly, Hollis placed the completed Koch folders and filings on George's desk. She would rather have waited until after she and Kelly visited with Ferris, but she had already pushed George's patience to the brink. Any new findings would have to be considered after probate. A periodic pang of guilt would creep into her consciousness when she thought about her heavy-handedness with Kelly. It was clear that Margaret Koch's story ended at the Eastbrook Hills.

She pulled out the thin file she'd pulled from Cathy's window-sill cache. Her eyes settled on the receipts and slips of paper. With the first deposition out of the way, she was anxious to provide Mark with more pointed questions for the upcoming ones. He had scheduled the next deposition for tomorrow. There was a chance these random pieces of paper held background she could use.

Hollis spread out the items on her desk. There had to be a reason why Cathy had them hidden, but none was immediately apparent. She and Mark had already gone through the stack of notes and found little of interest. What remained didn't mean much, either. In order to leave nothing to chance, Hollis volunteered to follow up with the last of the papers.

Three articles, a phone message and a receipt for $730.00.

The first article was about Fields being named to the list of Top Ten Bay Area Men of the Year two years ago. The short Associated Press article provided only a brief paragraph about each man's philanthropic activities and went on to discuss their personal struggles and ascents to fame. Fields' background and legacy were heads above the others. But there was nothing here Hollis didn't already know from Cathy's other notes.

The second article was about a 1990 embezzlement case in San Francisco. A Harold Roemer, head of a local accounting firm, was caught skimming thousands of dollars from a civic theater group. That in itself was bad enough, but he had run up debts in the tens of thousands and the theater faced ruin. She noted there was nothing about Fields in the article. It was likely research for another story altogether. Cathy had circled his name in red. Roemer should be out of prison by now.

The third article, written several weeks ago, was more of a promotion piece for the upcoming grand opening of a Napa winery. The owners, an old valley family, had dedicated all the proceeds for that year toward a drug rehabilitation center in memory of their son who had overdosed. Cathy had drawn a red star in the upper corner.

She was stumped. Except for the Top Ten article, she couldn't see a Fields' connection.

Still, she had to start somewhere, and all she needed was one tiny break. She reread the grand opening promotion. The name and contact information for the winery was at the end of the article.

Pulling out her phone, she punched in a number. She was transferred only once before she reached the owner.

"Mrs. Mueller, I understand you and your husband own a winery in Napa County. Your name has come up as a person who may have information that could be of help to my law firm in an ongoing case. Would it be possible for me to come out and speak with you for a few minutes?"

"Who is your client?" The woman's voice sounded cultured and imperious.

"I'm sorry, I'm not at liberty to say." Hollis knew that there would be no meeting if the Mueller's found out that she represented a tabloid like *Transformation* magazine. "I could be there any time tomorrow if you could fit me into your schedule."

"Who gave you our name?"

Bright lady.

"Catherine Briscoe. Do you know her?"

She paused, "Why, I think so. I should say we know of her. She tried to reach me several times, but we kept missing each other. She and I never talked. My husband did finally meet with her. I'm not sure I could be of any help to you."

Hollis couldn't let the opening close. "I promise I won't take up much of your time."

"Well, all right. Come tomorrow at one. But don't be surprised if it's a wasted trip."

A Tuesday meeting would barely give her time to get to Napa after the Meyer deposition in the morning, but it would give her time this afternoon to work on a few client files and search the Internet for any additional info on the Muellers.

THE NEXT MORNING Hollis was confident she had prepped Mark as well as she could. They waited in the *Transformation* conference room for the arrival of Fields' attorney and the soon to be deposed Phyllis Meyer, Fields' accountant.

There was a light tap at the door before it opened to a

smartly dressed woman and a different attorney from the first deposition. Hollis retrieved the court reporter from the adjacent room.

The attorney, who kept winking at Hollis, pushed a piece of paper across the table. "To save time, we'll stipulate that Miss Meyer approves all our annual reports for our northern California organizations."

Mark briefly scanned the sheet. "We acknowledge your stipulation. Shall we get started?" He turned to the sheet of questions Hollis had prepared. "Are you familiar with the Open Wings organization?"

Phyllis Meyer tossed her hair back. "Yes."

"Do you know Lilia Martini, the director?"

"Yes."

"Can you explain why Ms. Martini would say that she is the only employee working for Open Wings?"

"Because she is."

They waited for her to continue, but there was nothing else.

Mark spoke. "Then perhaps you can explain why Open Wings' annual report shows five employees and yet, as you agree, only Ms. Martini works there."

Phyllis Meyer could be a candidate for an Oscar. Hollis watched as she raised herself in the chair and leaned over the table toward them, her French manicured nails tapping against the high polished conference table.

"Full-time equivalents and overhead. We hire the homeless to put in just a few hours each week."

Hollis bit her tongue. Overhead. She realized the number shown in the annual report indicated a consolidation of employee work hours. Fields of Giving would charge back to each organization a portion of an administrative employee who supplied support. Each of those portions would be added together to come up with a whole number. While it might be misleading, it wasn't illegal.

Hollis gave Mark a look that he rightly interpreted.

"All right. Tell me how the funds come in to support Open Wings' operations"

"What do you mean?"

"Is there specific fundraising activity done just to serve Open Wings?"

"Yes." Meyer hesitated. "Can you just hold on a moment?" She whispered into Dapper Dan's ear. He whispered back. She nodded for Mark to continue.

"Okay, is the amount shown as income in the annual report accurate?"

She spoke slowly. "Yes and no. I posted what we knew at the time. Later, I had to make adjustments and submitted a correction to the Board."

Hollis was curious as to what aspect of that sentence required Meyer to confer with her lawyer. She knew Mark was making a notation to have the details about the corrections supplied.

He went on. "That actually is my next question. Who is on the Board for Open Wings?"

That made the cool Phyllis Meyer actually squirm in her seat.

"Myself and Hal ... I mean, Mr. Bartlett."

Bingo.

HOLLIS WAS RUNNING late. Wrapping up the deposition had taken longer than she anticipated. Mark said he would take care of briefing the *Transformation* legal team on the outcome. They should be pleased.

THE DRIVE TO the Mueller's home in Napa was a commuter's nightmare. As usual, the Highway 12 interchange was backed up and contributed an additional hour to what was already a forty minute drive. When she finally made it off the main highway she followed the GPS to a winding roadway and eventually to a mansion that resembled an ancient Italian villa. It was surrounded by acres of vineyards, and a large metal building rose in the distance that Hollis knew had to be where stored wine casks resided. She also took in the beauty of the rolling hills and the smell of ripening grapes that filled the air. Looking at the expanse of the front veranda and the pots of hanging geraniums and varicolored coleus plants, it was hard to believe the place had only been here seven months.

Her background search revealed that Arlo and Summer Mueller had purchased the property two years ago and immediately commissioned a famous San Francisco architect to design and build their estate. It had taken over a year to complete. Arlo was an entertainment producer and Summer a trophy wife with old money. They could afford to wait.

"Miss Morgan? Summer Mueller. Let's go into the sunroom and talk." Summer was wearing white crop pants and a Mexican style smock. Her perfectly coiffed white hair was pulled back into a small floral pin. Hollis guessed her to be in her early fifties.

They settled into wicker chairs after Hollis turned down her offer of iced tea or any other beverage to her liking.

"So ....?" Summer tilted her head.

"Yes, I understand you're getting ready for a grand opening. You must be very excited." Hollis looked around the room with admiration.

It was the right thing to say. Summer smiled with pride. "It's our life's dream. It was delayed, but now everything is going well."

Hollis mentally scrambled to come up with conversation that would elicit answers opening the way to asking further questions.

"How long ago did Catherine Briscoe contact you?"

Summer's lips drew into a tight line. "Before we go there, what is it exactly that you want from me? What information could I possibly have that could help you?"

Hollis said, "Catherine Briscoe was killed three weeks ago." Summer took a sharp intake of breath.

Hollis continued, "At the time she was in the midst of a lawsuit. We're representing her employer, who still wants to defend her work."

"How awful, but I don't see how I can help. To answer your question, she last contacted me on July eighteenth. I remember because our daughter got married the next day and I was frantic with last minute errands."

"What did she say?"

"Well, like I told you on the phone, I never did speak with her directly. She left a message asking to meet with us. After meeting with her the first time, Arlo was never going to meet with her again, so I knew it would just be me."

"Did she say what she wanted?"

Summer wrinkled her brow. "She wanted to talk about our opening. Well, not the grand opening ceremony but how we got started, our story. About Arlo's projects and my charity work."

"Did she say anything else?"

"No."

"Did she mention Dorian Fields at all?"

"Dorian, no, why would she?"

"Do you know Mr. Fields?"

"Of course we do. We work on several fundraising events and dinners together. We've been to his house, and he to ours."

Finally, a connection.

"Interesting. I'm trying to see the angle that would make your story of interest to *Transformation*."

Summer's face drained of color.

Hollis leaned forward. "You didn't know Briscoe worked for *Transformation*?"

Summer shook her head. "I think you should go."

"Mrs. Mueller, I'm just trying to retrace Cathy's steps. Can you just tell me if you had any contact with one of Fields' nonprofits—"

"I'm not going to answer any more of your questions and I'm

going to have to ask you to leave." She walked over and stood at the front door.

"I'm sorry, I didn't mean to upset you." Hollis picked up her purse and followed her to the entry.

"Somehow I don't believe you. Goodbye."

AT A DEAD stop in five o'clock traffic, Hollis saved a little time by using hands-free to call Mark with an update.

"So Mueller kicked you out?" he said.

"As soon as she heard that I was working with *Transformation*, I was *persona non grata*." Hollis slowly eased into the far right lane, taking the off ramp. She would take the side streets over the crowded highway. "Actually I don't blame her; I'd do the same. But the more interesting point was the Mueller connection to Fields."

"Yeah, but there're probably a lot of rich people who know Dorian Fields."

"But we've got a link between Cathy, Fields, and the Muellers. There could be something there."

"Sounds thin to me, but I guess we can't be choosy."

By the time Hollis returned to the office, it was too late to do much else other than pack up and go home. There was a message from Kelly stating that something had come up with her job, and she wouldn't be able to get away to see her grandfather this week. She would call Hollis back to make an arrangement for next week.

Hollis deleted the call in frustration.

THE NEXT MORNING, with Cathy's depositions behind her and the discovery of a possible Mueller/Fields connection, Hollis was feeling more confident. She didn't plan to be in the office so early but with the meeting with Ferris being put off for a week, she had time to catch her breath and work through some of the other case files George had left for her. Hollis spent the next two hours emptying her in-basket. By mid-morning she was completely caught up.

George came by and nodded approval when he saw the filings ready for his signature.

"I was a little worried that the Briscoe case was taking up too much of your time." He adjusted his glasses on his nose.

"It's nothing I can't handle." Hollis handed him the stack of large folders.

She gave him a slight wave as he left.

Closing the door to her office, she reached for Cathy's thin file containing the phone message slip. The yellow slip gave no indication who the message was for, just a time—2:35—a date, and the name and number for a Joe Morton. Hollis picked up her phone and tapped in the number.

"Morton's Photography, Amber speaking, can I help you?"

"I was trying to reach Joe Morton. Is he there?"

"Nah, he's out doing a shoot. Can I have him call you back?"

"When do you expect him to return?"

"He'll be back before lunch." The girl on the other end popped a wad of gum. "He's got an appointment."

"Okay, thanks. I'll come by before then." Hollis clicked off the phone. Why would Cathy take on a photographer for work assignments when *Transformation's* extensive freelance camera pros were some of the best paid in the industry?

Interesting.

MORTON PHOTOGRAPHY WAS a store front business located in a strip mall on the outskirts of downtown Oakland. Hollis parked in front of the bakery next door. A little bell tinkled as she pushed open the glass door. The room was split in half by a long, chest-high white counter. Poster-sized photos of brides and infants in a myriad of poses lined the walls. She caught a movement from behind the counter, and her eyes came to rest on a young girl—likely Amber—reading a paperback.

"Excuse me. I called earlier. I'm here to see Mr. Morton?"

"You're his appointment? You're early."

"No, I'm the one who called. You told me he would be back before lunch."

"Oh, well he's not—"

The door opened and a tall man with bright red hair, a beard, and freckles entered carrying a tripod and a large camera bag on his back. If Santa had red hair he would look like Joe Morton.

"Hello, you're early." He walked over with his free hand extended and shook Hollis'. "Just give me a few minutes to set up and we can get started. Did you bring your suit?" He put his gear down on a table in the rear of the studio.

While Hollis was tempted to take the conversation further and see what kind of photography Joe Morton produced, this was not the time.

"I'm not your appointment, Mr. Morton. My name is Hollis Morgan. I work for a law firm representing Catherine Briscoe's employer. I hoped I could just talk to you for a few minutes."

While Hollis had always thought a book on how to keep a poker face could really sell out to a niche audience, it was clear even if such a book existed, Joe Morton would never be able to play cards. His face turned beet red, and his hands clenched and unclenched. Even Amber looked up at his silence.

"Shit." He looked past her, letting a long moment pass. "Shit. I couldn't believe it when I read about her death in the paper."

He moved quickly behind the counter and put his gear on a rear table.

"If I could just—"

"I can't talk to you now. I have a sitting." His voice was strained as he moved hastily to organize the shoot.

Hollis ventured, "Can I see you later, at a time that works for you?"

"I'm not talkin' to no lawyers. I told Cathy I would only speak with her. She made me promise that no one else would know." He shook his head. "She trusted me."

"Cathy was also my friend. We're trying to defend her work. I found your name among her things."

He ran his hand over his beard.

Morton peered over an appointment book on the counter and ran his fingers down the page. "Amber, go take an early lunch. I've got the studio covered."

Amber scrambled to get her purse. She smiled as she headed for the door.

Morton motioned for Hollis to follow him to a back room. The windowless room was painted a soft taupe. It was tastefully decorated with a long sofa and love seat, low lamp light and framed oversized landscape photography.

"I got somebody comin' in, so we don't have a lot of time." He directed her to a chair across from a large rosewood desk. "Talk."

Hollis quickly described the arrangements with *Transformation* and the attempts she and Mark were making to validate Cathy's research.

She finished, "I know you have an appointment soon. Perhaps we could meet later today? I've given you a lot to think about, and I'd like to hear how you fit into all of this."

The front door bell tinkled.

Morton, who had been silent the whole time, finally nodded. "All right, come back at closing. We'll talk then."

WHEN HOLLIS GOT back to her office, she noticed a middle-aged woman sitting in the firm's lobby. Dressed in a faded red overcoat she clutched a purse in her lap as if fearful of a pending snatch.

Tiffany nodded at Hollis to come over to the reception desk. "She's one of yours."

Hollis raised her eyebrows and walked over to the woman. "Hello, I understand you're waiting for me. Did we have an appointment?"

"No, no, I just hoped I could catch you. My name is Amy

Hyde. Joy told me about you." Hollis must have looked uncertain. "You spoke to her at Heaven's Praises."

She smiled. "Of course, yes, Ms. Hyde, why don't we go back to a conference room, where we can talk in private?"

Hollis walked her to a small meeting room.

"I don't want to take up much of your time. Joy told me not to bother you. Everybody knows you've been going around trying to find out if we're doing our jobs. But I got to tell you: I don't think you realize what these places mean."

Hollis shook her head, "I'm not checking up—"

"I know you have to say that, but hear me out. I was an alcoholic for twenty-two years. There was nothing I wouldn't do for a drink. Nothin'. My family, they're all alcoholics, too. I was in and out of rehab more times than I can count. No one could tell me anything." She swallowed. "Then one night I went to Heaven's Praises. It was cold and rainy, the shelters were full, and me and my bottle planned to stay the night in the rear doorway. It was protected, and I could move the garbage bin over to keep me dry."

Hollis tried to keep the surprise from showing on her face.

Amy Hyde continued, "Anyway, I had all my setup done for the night when this man came around the side and moved the bin back. He was dressed in jeans and a slicker. He looked down at me with eyes I had only seen before in church when I was a child, and he held out his hand."

Hollis found herself hanging on the woman's words.

"He said, 'I've been where you are. Leave your bottle and come with me.' And even though I was drunk as a skunk, I had one moment of clear thinking. I went with him. We walked through that moment with me holding his hand."

Amy pulled her sweater close. "That was twelve years ago, and that man was Dorian Fields, and I've never had another drop since."

Hollis struggled to find the words. "Ms. Hyde, you don't understand. I'm not trying to—"

"No, you don't understand." Amy pointed her finger. "There are people who talk about doing good things, and there are people who do good things. Mr. Fields is the most honorable man I know. We may not always get the work right at the centers—I know you probably heard that Richard had all the linens stolen at Fresh Start, but they wasn't stole they was miscounted. And we know when Marian—"

"Whoa, Ms. Hyde, there's no need to explain anything to me." Hollis leaned over. "I do understand what you are trying to tell me. The centers mean everything to people who have nothing." She reached over and touched her shoulder. "Thank you for sharing your story with me."

"It's not just a story, Miss Morgan," she snapped. "It's the truth."

From the window she watched Amy standing next to the bus stop. She refused to let Hollis pay for a cab. As much as Hollis wanted to dismiss her, she couldn't. At first she suspected that Bartlett or maybe Fields had sent her.

But deep inside, she knew the woman was telling the truth.

"COME ON IN the back," Joe said.

Hollis followed. The studio atmosphere, like its owner, seemed still and subdued. She sat in the same seat she had this morning.

"I still can't believe Cathy is dead." He squinted at his hands, folded on top of the desk. "We almost didn't work together, but we ended up being friends."

Hollis reminded herself to keep her impatience in check. "Mr. Morton, how did you know Cathy?"

"Call me Joe. I met Cathy about six months ago. She was doing some research on a story. I'd done some photos for a customer, and she found me through one of them."

"Why do you say you two didn't start out well?"

He ran his fingers through his hair. "When she first came to the studio, she wanted to know if I doctored pictures. She had one and she wanted to know if I could doctor it." He shook his head. "I didn't like the way that sounded, so I told her to leave and take her picture to somebody who didn't know better."

"What did she say?"

"She laughed." He looked into Hollis' eyes. "She said that I was the man for her."

Hollis wondered if Morton had had a crush on Cathy.

"Ah, it was a test."

Misery washed over his freckled face. He appeared crestfallen; then he sighed and went on. "I really liked her, she was a nice woman and …."

His voice drifted and Hollis looked away.

"Mr. Morton—"

"Joe, please."

"Joe." She smiled and lowered her voice. "Cathy and I went to law school together. We were good friends. I'm representing her employer in a matter that involves determining the validity of her research for an article. I found your name among her things and I was hoping you could help me."

He straightened in his seat and took a deep breath. "She needed me to blow up some old pictures. Seven of them. She gave them to me on a thumb drive, so all I had to do was digitally enhance them."

"Copies of copies?"

He nodded.

"Do you happen to remember what was on them?"

He closed one eye and focused on the past. "Four were of a group of men on the steps at a conference, or maybe it was a reception. One of a woman at an airport—I think she was waving goodbye. And there was one of a document. Looked like some kind of business letter or memo."

"What about the last one?"

He pursed his lips in a tight line. "Sorry, I can't remember."

"What about the logo on the letterhead? Can you remember what it was?"

He squinted again, "I don't know, maybe an anchor or a shield? I'm sorry."

Hollis shrugged. "You were very helpful. I really appreciate your talking with me." She stood.

He slapped his head. "Wait, there was one other thing, I almost forgot. About three weeks ago she called me again and asked me if I had done work taking photos of dead people."

"Dead people?"

"Yeah, you know, like in the morgue."

CHAPTER SEVENTEEN

~~~

THE NEXT MORNING Mark passed a file across his desk to Hollis.

She marked it off a list and placed it in her briefcase. "We're done here. I've got to get back to the office and finish up a couple of cases on my desk, but I want to brief you on my visit to the photography studio yesterday evening."

"Did Morton know anything?"

Hollis quickly ran through the conversation.

"Morgue?" He shook his head. "We've got our settlement hearing coming up fast. Do you think there's anything to these photos?"

Pressing her lips into a thin line, Hollis said, "I honestly don't know. Morton seemed legitimate, but he didn't appear to know anything about Fields. It may be that Cathy was using him for work on a future article. I just don't have enough information to know if the photos are relevant to our work."

"Why would she want a picture of a dead person?"

She shook her head. "I don't know. This is what drove me crazy about her. She was so secretive. She called me paranoid, but she was my mentor." Hollis played with her pen. "Suppose

Cathy found out that Fields was laundering money from the nonprofits into his own accounts. From what I could gather from her notes, she wanted to roll out his story over three issues, and had only turned in her first installment to *Transformation*. The nonprofits themselves might be genuinely innocent, because they had nothing to do with their own account books."

"And where do the dead people fit in?"

Hollis dragged her fingers through her hair, "Awrrrrh."

"I'm with you on that," Mark said. "We don't have much time to speculate. Let's leave the dead for now and focus on Open Wings. It's the center with the most unanswered questions."

"All right. I'll delve into Open Wings' operations. If we can find more holes in their story, we'll know where to start looking in the others." Hollis picked up her purse.

Mark frowned. "Don't be getting too far out there. It might be time to let the police know."

"Know what? That's just it. We don't *know* anything." Hollis put away the last of her pens and markers in her purse. "The interviews, the depositions, the random notes all say something, but what? We're overlooking it."

He nodded. "Yeah, other than uncovering some questionable business practices, we've been dancing around. Still, you seem to be ruffling a few feathers. You let me know if things are getting too … well, dangerous."

Hollis patted his hand. "Now, don't I always let you know—eventually?"

"Where are you headed now?"

"I've got to go back to the beginning, one more time through all the events, pieces of paper, and interview notes. I've got to find the missing connection."

HOLLIS CLOSED THE door to her office. Leaning back in her chair with eyes closed, she rocked back and forth. Instead of having no clues, she had too many. It was clear the non-profits were indeed running off "track"—in fact, it was too easy to see. It didn't take an investigative reporter to find the glaring discrepancies.

In the same vein, Gail Baylor's murder following close after Cathy's left the impression that the murderer was desperate, and whatever he or she faced was worth killing for … twice. If she followed that premise, her own break-in was a red flag, and she could be next on the victims list.

She picked up the receipt she had gotten from Cathy's apartment. It was orange, five by six inches, a standard form, much like any generic receipt you could get from Staples or any office supplies store. It had even less information than the phone message; at least the phone message had a number she could call. The receipt listed just the amount—seven-hundred and thirty dollars for consultant services.

"You look deep in thought, what's on your mind?" George plopped himself in her chair. "Worried about the exam? Scores still come out in November?"

"Some things never change." Hollis grimaced. "You know the California Bar; they make us suffer until the brink of a breakdown."

"Well, you appear to be holding up fine. Except, what has you so frustrated?"

Hollis shrugged. "Mark and I have pulled together as much as we can on Cathy's case, but the dots aren't connecting like we need them to."

She picked up the receipt. "For instance, Cathy left this behind in a research folder. It has some importance because she felt she had to hide it, but it tells me nothing."

"Let me see it."

She handed it over.

George examined it for a few minutes. "I can tell you a couple of additional things. First, the date is in March, but the reference number is 0012. Assuming they operate on a calendar year, it's a low volume business to only have twelve invoices in a three-month period."

Hollis nodded. "I'm impressed. What else?"

"Most importantly, the name of the company is at the bottom of the page." He pointed to the name in the lower left corner. It was in a tiny, embossed eight-point font.

"Let me see that." Hollis snatched the paper from his hand. George laughed.

"Templeton Group." Hollis laughed, too. "I'd hug you if it wasn't sexual harassment. Thank you. I didn't see either of those things. I'll check them out."

"You didn't see them because you've been working too hard. You need a break. You got the Koch matter settled; take an hour off for lunch."

"Very funny," she said. "I'll slow down once this matter with Cathy is resolved. And then I'll take some time off after I get my scores."

He got up. "I hope your life can wait for later."

SHE WAS BEING followed.

She had the same feeling when she left Mark's office the day before. Now the feeling was back, and stronger.

At first Hollis thought her peripheral vision was in too high a gear, but as she wandered through the mall, she sensed, more than saw, someone watching her. She pulled out her cellphone and held it up. It made an adequate mirror for seeing behind her. She whirled around.

"Vince?"

The young man walked slowly toward her, hands shoved in his jeans' front pockets, the gray hoodie pulled over his straggly brown hair. "Hey, Hollis."

"Are you following me?" Hollis pointed to a nearby bus bench then tugged his sleeve for him to come with her.

He allowed her to lead him. "Ah, just hanging out. I wasn't doing anythin'."

"So why follow me?"

"I didn't have nothin' to do. I saw you with Stephanie last week while I was waiting for my mom." His head was down, his words mumbled.

"How's your mom?"

"Oh, she's out now," he mumbled. "But she has to put more time in out-patient rehab."

Hollis felt a twinge of sadness. Her own family kept their distance from one another so they wouldn't have to get involved—or be inconvenienced. Then here was this youth whose care and involvement with his mother was so integral to his purpose in life.

She gave him a half-smile. "Still, why follow me? I'm not that fascinating. Shouldn't you be in school? Or at a job?"

"Nah, I dropped out of high school. I work at the Fast Stop at night, cleaning up."

She noticed his eyes would not meet hers.

"Are you still getting over the drugs?"

Vince's voice rose defiantly. "I ain't touched those in weeks. I'm clean. I didn't have to do no rehab. I did it all myself."

"I believe you." Hollis smiled back at curious onlookers and said in a low tone, "I'm glad."

For some reason she felt a tie to this young, ungainly youth. He reminded her of no one she knew, but she felt drawn to his story. It was the same with Margaret. The compelling stories of people who dig deep to change, grow in character and overcome odds were the ones she clung to, only she hadn't realized it until she found that picture in Cathy's condo. It gave her hope.

Chest heaving from emotion, Vince sought to regain a steady breath, all the while continuing to nod.

Hollis pointed toward the food court down the block. "Are you hungry? Want to get some lunch?"

He jerked his head up with a look of such amazement that Hollis was taken aback.

"What's wrong?" she asked.

"Nothin."

"Well, do you want to get something to eat, or not?"

"Yeah, okay."

They picked a table with facing seats. Hollis ordered salad and Vince, hamburger and fries. With his head held down, Vince responded to her attempts at conversation with one word responses.

Hollis took a sip of tea. "Now tell me why you were really following me."

Vince pulled his hood closer to his head. "'Cuz like I said, I didn't have anything to do. I saw you in the parking lot and I just followed you."

A slightly different story than the one he had given her before, but Hollis held her tongue. "Did you follow me yesterday?"

Vince's hood could not hide the flame of red that crept up his neck to his face. "Nah, that wasn't me."

He's lying.

"So, you left high school." Hollis picked at her roll. "Are you working on getting your GED?"

Vince jerked his head up. "What? No, well, maybe. I don't need school no more. They kicked me out for being a truant. But I had to take care of my mom."

Hollis sighed as she nibbled on a cucumber slice. "What was your favorite class?"

"I don't know, maybe … maybe," he stopped eating and looked past her, "maybe history."

His answer surprised her. It was not a subject she would have thought he'd be interested in. "History? Why history?"

"Because it's all over. Everybody knows how things turn out. I like to know how things turn out."

She tapped her lips lightly with her fingers. "I never thought of it that way, but I see your point."

They each took another few bites in silence.

When Hollis finished her salad, she said, "You know, if you wanted, I could get you the forms to go to continuation school and see about that GED."

Vince frowned.

Hollis held up her hand. "It's no big deal. I can download forms off my computer at work."

With his head still down, Vince pushed back his hood. With his tow head uncovered, he looked even younger. "I'm not ready. I gotta help my mom get through out-patient rehab."

Hollis nodded. "Okay, okay. Let me know if you're ever interested."

"I'm not ready," Vince repeated.

She checked the time on her cellphone. "I've got to get back to the office, but you stay here and finish."

"Er ... thank you for buyin' me lunch."

Hollis smiled. "It was my pleasure, Vince."

"Why did you?" Vince hesitated. "Why did you spend time with me?"

Hollis raised her eyebrows and shrugged, remembering her friend, Cathy.

"Because I could."

HOLLIS PUT HER purse in the lower desk drawer and pulled out the Templeton invoice. George was right. The side trip to the mall and lunch with Vince was just the break she needed.

She Googled the Templeton Group and watched the screen load with ads, sidebars, and quotes—none of which actually indicated what business they were in. The icons at the top were generic descriptions. After hitting the "Contact Us" tab, she took down the San Francisco phone number and punched it in.

"Yes, my name is Miss Hollis Morgan. Can I speak with someone in your accounting department regarding an invoice we received?"

A few seconds later the same voice who answered the phone came back on line. "This is Nancy, how can I help you?"

Hollis smiled. "We have your invoice number 0012. I had a couple of questions."

"Oh, yes, I know that invoice. What's the problem? I know we cut the check about three weeks ago. You should have received payment."

Hollis bit her lip. The invoice was for Templeton to pay, not owed to Templeton. She scrambled to recast her inquiry. "Oh, no problem with your check, I'm calling to see if our service was satisfactory."

"I suppose." Nancy paused. "Mr. Mueller handles that account. I don't know anything about it. I just cut the checks."

"Arlo Mueller?"

To say her voice had cooled with suspicion would have been an understatement. "Yes."

"Is Mr. Mueller available?"

"No, you'll have to call back."

Click.

That went well.

Hollis grabbed her purse and keys. Let's see what they would do if she showed up on their doorstep.

But by the time she pulled her car up to the three-story office building on San Lorenzo Boulevard, she was having second thoughts. The ground floor windows were shuttered closed. The second floor windows were boarded over. Surprisingly, the entry door opened easily into a stark, dilapidated, and empty lobby. On a rear wall, a tarnished metal case contained a directory. Templeton was on the top floor.

The elevator lumbered noisily and eventually opened up to a hallway containing four doors, one of which was glass and touted Templeton Industries, Inc.

Hollis handed her card to the woman sitting at a long narrow table that substituted for the receptionist desk. "I called earlier. But I'm willing to wait for Mr. Mueller until he returns." She took a seat on an uncomfortable green metal chair.

"I told you on the phone he wasn't available." Nancy, who appeared to be in her seventies, wore her blond waist-length hair styled like Alice in Wonderland—held back with a black headband. However the skin on her face—unlike Alice's—fell into small folds with a dot of red on each cheek.

Hollis ignored the glare. "Yes, well, I was in the neighborhood."

Nancy pushed a button on the phone, murmured in response and stood. "Please stay in your seat. I'll inform the manager, Mr. Green, you're here."

Stay in my seat.

It wasn't three minutes until Nancy shuffled back to her desk.

"Mr. Green will see you once he completes a phone call."

Hollis beamed a fake smile.

It took another five minutes for Green to appear. He strode into the room with both hands in his pockets. Rumpled, balding, and middle-aged, he wore a dark brown belted sweater and an open-collared, peach-colored shirt.

"Miss Morgan, if you'd come this way." He pointed toward an office.

"Mr. Green, my law firm is working with *Transformation* in defending an article written by Catherine Briscoe, We—"

"My assistant gave me your card." They were in his office, and she took the chair he offered. "I don't see how I can help you."

"I'd like to show you a photo of Cathy. It might trigger your memory. I also wanted to talk to you about a receipt."

"Yes, I understand that from my assistant. She told you that we don't know a Catherine Briscoe. She also told you our accountant paid the invoice."

"Mr. Green, please, our client died a couple of weeks ago. We discovered this receipt among her papers. We just need to know what the service provided was."

He was silent. Then, "Who did you say you were again?"

Hollis paused. "My last name is Morgan, and I'm with Dodson Dodson & Doyle. As I mentioned earlier, my firm is working with the McClouds law firm and their attorney, Mark Haddan. We're defending *Transformation* magazine in a lawsuit filed by Dorian Fields." She leaned in. "Could you tell me your connection to Dorian Fields?"

"Fields, the philanthropist? I don't know Dorian Fields at all."

She sank back and reached inside her purse, thrusting the photo across to him.

"This is a picture of Catherine Briscoe. Do you recognize her?"

He gave it a brief glance. "Pretty lady, but no, I don't remember her."

She was running out of threads. "Mr. Green, a few moments ago you reacted when I asked what service you provide. Now, I'm just a paralegal. I'm only interested in an article that Cathy wrote for *Transformation*, so you can tell me about your business, and I will not reveal any confidences."

"Look, I don't know you and I don't know Fields and I sure don't know Briscoe, but I do know that the receipt you showed my assistant is for the association fees."

Hollis frowned. "Association fees? What association?" She mentally checked possibilities but nothing clicked. This wasn't making any sense.

"Your question says it all. Clearly, you're not privy to the confidential transaction involved here. I need to get back to work. So I'm going to have to show you out."

"I'm sorry if I said something to bother you."

"You can tell him for me that he doesn't have to send messengers by to see if I'm playing by the rules. I'm a big boy and I don't appreciate it."

He escorted Hollis to the door, a firm grip on her elbow.

Who's *him*?

CHAPTER EIGHTEEN

~~~

HOLLIS SAT ON the edge of Cathy's bed. She was grateful that Friday was a slow day in the office. She was able to leave and come here to finish packing her friend's personal belongings. She didn't mind; it helped her to say goodbye. Tomorrow she would call Mrs. Briscoe and tell her that her daughter's things were ready for the movers.

She sniffed. The air still held the faint scent of Cathy's favorite Jo Malone fragrance, Red Roses.

"Talk to me, Cathy," she said in a whisper. "What were you up to? What do all these pieces mean? There's Templeton Industries and Joe Morgan's photos, the Muellers. What are the connections? What's the link? What does one have to do with the other? Fields' nonprofits were a little sloppy on the management side, but they appear to be doing good works. What did you see that I don't?"

Engrossed in thought, she was brought swiftly back to the present by the click of the lock on the front door. Her body stiffened, and moments later Hollis looked up into the startled eyes of a medium-height, brown-haired, good-looking guy with beard stubble. He carried a large backpack.

"Who are you?" he said, looking around the room.

Hollis jumped up from the bed and stepped backward to the nightstand, ready to defend herself with the lamp, if necessary. Adrenaline pumped through her body.

"Who are you?" Her hand felt for the lamp base.

"Where's Cathy?" he said. He dumped his gear inside the room but didn't approach any closer. "What's going on?"

Hollis took a breath. He knew Cathy. He seemed okay, and he had entered with a key. Her heartbeat was slowing back down. She didn't feel threatened, but she wasn't comfortable either.

"I'm a friend of Cathy's." She motioned toward the door. "Let's go sit in the living room."

He looked doubtful and held his ground until Hollis pointed him into the next room.

She took a seat in the single chair by the window. He sat on the sofa, on the other side of the room.

"My name is Hollis Morgan. Did Cathy ever mention me?"

He shook his head.

"What's your name, and how do you know Cathy?"

"Michael, Michael Carver. Cathy and I are … together." He leaned forward and gripped his knees with his hands. "What's going on?"

Hollis sensed he was the real deal and wished she were anywhere else but there.

"She never mentioned you to me, either, Michael. I'm afraid Cathy was killed about three weeks ago." He looked as if she had punched him. "I'm sorry."

He stood and ran his hand over his brow.

"I don't understand. How was she killed?"

Hollis debated whether to give him the full version or to parcel it out in segments. She opted for the former. His reactions went from disbelief to anger to pain. She finished.

"The funeral was two weeks ago. I'm sorry."

He was silent, his disbelief visible.

"She was always so independent. She was funny and cre—" His words seemed addressed more to himself than to Hollis. He slumped back onto the sofa.

"I know. We were good friends." Hollis paused then said, "Cathy never told me about you. Where've you been?"

He looked at her with red-rimmed eyes. "Camping. I've been camping in Yosemite. I go every year for one month." He put his head in his hands. "I didn't think she wanted me to go this time. We hadn't known each other that long. She didn't say anything, but … I knew …. If I hadn't …."

Hollis came over and sat near him. "Don't even go there. I've known Cathy for years. She wasn't one for hand-holding or holding hands. At least not in a non-romantic way."

He gave her a half-smile and wiped at his eyes. "Why do you think she didn't tell you about me?"

Her first thought was that Cathy didn't consider him a real contender. But it could also be that she wanted to be sure. She had given the guy her key, after all.

"She was funny that way," Hollis said. "I've been pretty busy these past months studying for the bar, so we hadn't seen each other for a while. The night she died she came by my condo. That was the first time we'd seen each other in months." Hollis reflected on the truth of her words. "She was preoccupied with her work and I was focused on the bar. We really hadn't talked."

"You a lawyer?"

"I'll know in a few weeks." Hollis smiled. "What do you do?"

"Risk management. I sell insurance."

Hollis' stomach fluttered. Insurance was a field she was intimately familiar with. She flashed forward to memories of her ex, who had no qualms about setting her up to take a prison term for his insurance fraud scheme.

Michael said, "What's the matter? You look like … like you just remembered something."

She shook her head. "No, I'm okay." She went to the kitchen for a glass of water. She pointed to a glass. "You?"

"No, thanks." He cleared his throat. "Do they have the guy who did it?"

"No. The police don't even know if it is a guy."

He looked pensive. "Is there a motive? Why was she killed?"

Hollis shrugged. "The police don't know. But I think it was because she was writing an inflammatory article about a celebrity, a celebrity who couldn't afford to have his reputation sullied."

"You mean Dorian Fields?"

Hollis turned to him in surprise. "You know about the article?"

"Yeah, sure."

He looked taken aback at Hollis' reaction. "Cathy gave me an early draft to read. I took it along on my trip, but to be honest I didn't get a chance to read it."

"You have the copy with you?"

"Yeah, sure."

"Michael, my law firm is working with *Transformation* magazine to defend Cathy's article against a lawsuit. Fields filed for libel." Hollis reached into her tote for a pen and notepaper. "Most of her research and draft copies were taken by her killer. We've been trying to piece together her findings. I'd like to compare the article Cathy gave you with the one we got from *Transformation,* just in case they're different versions."

He frowned. "The killer took all her research. Then the police must think Fields did it."

Hollis ran her hands through her hair. "Well, at first they thought she committed suicide because of the lawsuit. But in light of subsequent events, murder appears more plausible. Fields is very powerful. I think the police are looking at any and everyone else first."

"Before I left, she was really jazzed about something new she'd found out."

"Hmmm. Like I said, she came to see me that night be— before ... really excited, but depressed too." Hollis paced around the room. "We almost had our hands on the notes she left with her assistant, but then she was killed too."

That brought Michael to his feet. "Gail was killed, too?"

"You knew her?"

"Gail was a little strange, but she was okay. Cathy relied on her to keep all her materials straight and transcribe her drafts." He shook his head. "When I met Cathy for lunch, Gail would usually call and interrupt with some question she was worried over."

"Gail was going to give me a thumb drive with Cathy's research," Hollis said, "but she was killed before she was able to pass it on."

Michael stood a moment with his own thoughts. Looking up, he said, "This Fields guy must be pretty bad."

"You know, he's like Dr. Jekyll and Mr. Hyde. The people he reaches out to think he walks on water, the business world holds him up as an angel, and yet I'm coming to believe he is cold-blooded enough to kill to protect his image."

He rubbed the back of his neck and closed his eyes. "I need to think. I can't stay. I need to take this all in."

Hollis nodded. "Can I get your copy of Cathy's article? It might me help figure out her early research sources."

"Sure, I guess." He reached into his backpack and stopped. "Do you have a business card?"

Hollis nodded in understanding. "Yes, here." Pushing aside a gnawing feeling, she handed it to him. "Can I get your contact information?"

He handed over the article then scribbled on a slip of paper. He left with the promise to keep in touch.

Hollis hadn't had the heart to tell him about Cathy's ninety-day guy policy. As Michael told his story, Hollis glanced around the room. There were no pictures of them together. Cathy didn't like seeing a guy more than ninety days in a row. More time than that, and she felt she couldn't get rid of them. She'd been burned badly once, so she had no qualms about sending a guy packing, changing her locks, and not looking back.

Hollis sat on the sofa and read through this early version of the article. It wasn't long but it was clearly written. The tone was a little mean-spirited, and it raised questions rather than pointed to hard facts. She laid the pages aside and stared off into space. It was more accusatory than the final version.

Without proof, the article could be considered libelous all right.

BACK AT THE office, Hollis had only one message on her phone. It was from Kelly confirming that she had completed her job assignments and could meet with Hollis at the residence home on Wednesday. Hollis sighed. The court clerk had already given her a probate hearing date on Friday. The Koch estate would be processed without heirs.

She gave a wave goodnight as she passed through the lobby.

"Doing anything special this weekend?" Tiffany asked Hollis, while straightening papers on top of the reception desk.

I wish.

Hollis paused, shifting her purse and jacket. "No, just housework. See you Monday."

She heard her phone ringing as she stuck her key in the condo door lock. She rushed over but the answering machine caught the call.

It was Faber.

"… wondering if you might be interested in dinner and maybe a movie tomorrow night. I know it's short—"

She snatched up the phone. "Hi."

"Hi," he said. "I know it's short notice, but I wanted to invite you to dinner and a movie tomorrow."

"As it happens, I'm free, thank you." Hollis was glad he couldn't see her punching the air with her fist.

"Good. How about I pick you up at five? Would you rather see the movie first or have dinner?"

"Surprise me."

Hollis leaned against the wall, a broad smile spreading across her face. She punched the buttons on the phone.

"Stephanie, John Faber just invited me out. What should I wear?"

"Boy, your stars must be aligned. Second date in less than a month. What happened to the other guy, what was his name?"

"Brad. He's nice, but he's friend material. I don't think Brad and I have the same values for a long-term, or even a short-term relationship. I love reading; he got stuck on *Old Yeller* in grade school and never finished another book. He loves football—did I say 'loves?' More like is obsessed with it. After fifteen minutes we have nothing to talk about. Besides, he's on a business trip." She threw up her arms. "What should I wear?"

"Jeans, your maroon top, and that fringed shawl I gave you for your birthday."

"That's a great idea. Thanks." She was already picturing herself looking stylish, but fun. "How are you doing? We need to have another lunch. You won't believe who followed me to the mall the other day and half gave me a heart attack."

"Who, the other guy?"

"No, I told you his name is Brad and he's out of town. It was Vince."

"Who's Vince?"

"That kid you helped get into the shelter program. He's still clean, not so jumpy. There's a question about his mom, but he seems to be on his way back."

"I'd write him off. There's too much going against him."

Hollis murmured, "Yeah, I guess."

"What?"

"I guess I still believe in second chances."

FABER DABBED BUTTER on a piece of roll. "What did you think of the movie?"

"To be honest, it has been so long since I've had time to go to a movie, I was blown away by the 3-D technology." Hollis laughed at herself.

Faber gave a hearty laugh. "Yeah, I can understand."

She took a sip of wine. "Tell me about you. Were you raised in California?"

"Born and raised in the Bay Area, went to St. Mary's, and entered the academy when I was twenty-three." He looked down, and then into her eyes. "I was married two years when my wife died having a still birth."

"Oh, I'm so sorry," Hollis said. "How long ago?"

"Seven years." John looked reflectively at his hands. "I'd just gotten a job on the force." Then he seemed to notice Hollis' discomfort. "Hey, I've healed. There's just a little scar tissue there, but I've come to grips with it." He took a sip. "And you?"

"Well, you already know everything about me."

"I know what the State of California says, but I don't know you."

"Typical dysfunctional childhood." She picked at the remnants of her salad. "My dream is to pass the bar and become a probate attorney."

"Why that area of the law?"

"It's narrow, it's interesting, and it's rewarding," she said. "It's clear and straightforward."

"It's mostly paperwork."

Hollis smiled, "Yeah, I know."

He nodded. "What about your family—any brothers, sisters?"

"One each. What about you?"

"Only child. I was adopted."

"I wanted to be adopted," she said wistfully.

He looked at her curiously.

She shook her head. "Just kidding. That slipped out. Ignore me."

They chatted companionably through the rest of dinner. Hollis found herself not wanting dessert to arrive to mark the end of a great evening.

John leaned back in his chair. "Hey, I'm going up the coast

next Friday to pick up a table for a friend. Would you like to take a drive with me? We could make a day of it."

Hollis looked dejected. "I can't. Cathy's hearing starts next Friday."

"Maybe another time," he said. "Right now I have Thursdays and Fridays off, but next month I'll get my weekends back." He paused. "How's the case going? Did she have the goods on Fields?"

"We think so. How's Cavanaugh doing? I haven't heard from him lately. Does he have any clues to Cathy's killer?"

"Nah, but he's on it. No detective wants to have a double murder on his plate. He's doing his job."

"Mark and I have been coming up with some curious material, but without Cathy here to tell us what it all means, we can't tell if there is anything we should run with. It's pretty maddening."

John picked up his cup of coffee. "You haven't had any other burglaries or intrusions, have you?"

"No, and remember: whoever it was knew it was my place."

"Yeah, that's the part that worries me."

"But evidently Cavanaugh still doesn't think it's Fields."

"You've got to admit, Fields wouldn't be that stupid. Besides, he's taking his side to court. There's no one else."

"What about the boyfriend?"

John eyes narrowed. "What boyfriend?"

"A Michael Carver," she said. "I was in Cathy's apartment day before yesterday when he entered with a key."

"You tell Cavanaugh?"

Hollis looked contrite. "No, I honestly didn't think to, until now. I guess I really should. Carver said he was out of town; maybe he went to see the police on his own."

He looked at her skeptically. "Remember to call Cavanaugh tomorrow. You don't want to be seen as concealing information."

She rose. "I'm not concealing information. Anyway, Cavanaugh doesn't really think it was Fields. Why should I do his job for him? My job is to clear her name. She at least deserves that."

John lowered his voice. "Did you learn anything from this boyfriend?"

Hollis took the hint and lowered her voice, too. "He didn't know she was dead. He'd gone camping with some friends a few weeks ago. He's been seeing Cathy for a couple of months before that. Which I should say is about her maximum term for a boyfriend. He did have a copy of her article, not the one we are defending but an earlier draft. That's all I know."

"Well, from what I can tell," he said, reaching for her hand, "I would definitely want you for my friend."

Hollis looked into his smiling face and felt warmth in her cheeks. "Thank you."

"What say we go listen to some music at Cleo's?"

"You might be a night owl, but I'm not." She smiled. "I'm beat. I'll have to take a rain check if you're giving them out."

"For you, there's an unlimited supply," he murmured.

They were both silent for a moment. Then he leaned over and kissed her gently on the lips.

And Hollis kissed him back.

CHAPTER NINETEEN

❧

HOLLIS SAT IN the lobby of the senior facility waiting for Kelly to emerge from her grandfather's room. She was frustrated at having to wait the additional week, but with the work needed for Cathy's research, the time had passed quickly. Fortunately Kelly was true to her word.

Since it was Social Wednesday, it took a little longer for visitors to check in and get past the front desk. They agreed that Kelly had a better chance of convincing Eric Ferris to speak to Hollis if she went to him first by herself.

She held her breath when she saw Kelly walking quickly toward her.

"After some arm twisting, he agreed to see you, but he's not happy about it. He says he's only doing it for me." Kelly stood by the hallway entrance waiting for Hollis to gather her things.

"I won't be long with him." Hollis walked behind her.

They turned onto a long corridor. About halfway to the end, Kelly stopped and pointed to the room. Hollis entered first.

"Mr. Ferris, I'm sorry about the ... my last visit."

Eric Ferris was dressed in slacks, shirt, and a sweater. He sat propped up on several pillows in a large lounger. Off to the side was a standing tray containing a small blue plastic pitcher of water and plastic cups. Hollis took a chair near the stand. Kelly sat opposite her, next to her grandfather, holding his hand.

Ferris growled, "My granddaughter insisted that I meet with you. I'm listening."

"This is the letter I wanted to give you from Margaret Koch. We found it among her belongings. She died in June."

Hollis held out the yellowed envelope, but Eric didn't reach for it. After a moment, when it was clear he had no intention of taking it, she put it back in her purse.

A small smile came to his lips. "Well, the bitch is finally dead."

Hollis wasn't surprised by the words so much as his vehemence. "The letter is addressed to you, but it was returned to sender. We only opened it to see if could affect her estate." She paused. "I … we wanted to make sure you—"

"She have any kids?"

"No, in her letter she—"

"I'm not reading any damn letter from her." He struggled to sit taller in the lounger, making it clear to Kelly, who now stood, that he didn't want assistance. He pulled the tray to him and shakily poured himself some water.

"Maybe I should go," Hollis said.

"No, you're going to hear this. I don't want you hounding me again." He stared out the window. "She thought she could buy me, the slut." He slammed the top of the chair arms with his fists. "We might as well have killed Charles. 'Cept he did it to himself."

Hollis looked over at Kelly, who seemed mesmerized.

Hollis frowned. "I … I don't understand. George Ferris killed himself?"

"I'm sorry, sweetheart." He turned to Kelly.

"Granddad, why?"

Eric's withered face turned away, but not before Hollis saw his eyes fill with tears. "I thought she did it. I could've sworn I saw her push him. All those years in jail ...."

Hollis stood next to him. "But how did you—"

"I found his note when I got out of prison." He wiped his eyes with the back of his hand. "He couldn't handle being a cripple and being betrayed by his brother with the love of his life."

Hollis was silent.

"So you see, young woman, take yourself and that letter, mind your own business, and leave me alone. There is nothing she could say that I would want to hear." His reddened eyes seemed to look through her.

Hollis picked up the letter and rose. "I'm so sorry. I ... I thought if you read her letter to you and ... you would know how Margaret felt ... from her point of view .... I didn't mean to upset you.

"Upset me?" He leaned closer and Hollis took a step backward. "She died last June, I ... I died fifty years ago."

Hollis nodded and stood to leave the room with tears forming in her own eyes. She looked back and saw Kelly holding her grandfather in her arms.

Kelly looked up at her. She mouthed that she would see her tomorrow. Hollis closed the door as Ferris' body seemed to collapse on itself and he broke into shaking sobs.

STILL RATTLED AFTER the visit with Eric Ferris, Hollis drove on autopilot to Mark's office. As they prepped for the *Transformation* meeting, he must have sensed not to engage her in conversation, other than giving her a quick overview of how they would proceed. Now they waited for Carl Devi and his legal team to appear. It was the last strategy meeting before the initial hearing. She tried to concentrate on the task at hand, putting the morning with Eric Ferris out of her mind.

Mark reshuffled papers and lined up his pens beside his legal pad.

"You ready?"

Hollis shrugged. "Yeah, I'm fine. We've got evidence that

Fields' centers were poorly run and that fact alone raises concern over the use of donations and contributions. There is no dispute; Cathy's allegations were grounded in fact."

Mark rubbed his forehead. "Yeah, but we don't have indisputable proof of mishandled funding."

Hollis didn't try to hide her concern.

"We need to face facts too," Mark said. "You should try sitting down with *Transformation*'s legal team to defend Cathy's words." He took off his glasses. "I think *Transformation* wants to settle."

Before Hollis could respond, Devi and his three attorneys entered. Introductions cycled around.

"Well, good morning, Hollis, we haven't seen you for a while, but I understand you've been working diligently behind the scenes."

"Very diligently. We'll be ready for the hearing."

"Then you must have learned something new that Mark hasn't shared with us," Devi said. "What is it?"

Mark interjected, "Let's go over our case. I think you'll see we have a good position."

Devi reached into the briefing folder Mark had passed him. "Good, but not strong?"

Mark didn't answer. Instead he began his oral argument.

Hollis looked around the room and noticed that Mark had the attention of everyone except Carl Devi. Devi was staring at her. When their eyes met, he looked away.

Twenty minutes later, Mark stopped speaking and asked for questions.

At the very least the attorneys looked skeptical.

The youngest one, an Asian woman, spoke first. "Mr. Haddan, I grant that you have framed a very difficult case in the most defensive manner possible. You've done a lot of work."

"But?"

The attorney sitting next to Devi adjusted his green plaid

tie and said, "But, where are the facts that support Briscoe's statements? She accuses him of laundering donations for his own use. Show me a fact, not a clue, not a contention, but a cold fact that says Dorian Fields defrauded contributors."

Mark tossed his pen onto the table.

"I could bullshit you, but this is as close as we could get."

"I thought so," Devi said. "Let's put an end to this." He turned to the attorney at the end of the table. "Float a settlement number and—"

"Wait," Hollis interrupted. "Okay, maybe we can't nail him on laundering, but there is still a lot of dirt that Cathy dug up. We can still use it. Doctored annual reports, misleading non-profit organizations and inflated expenses all add up, if not to something illegal then something that flies in the face of full disclosure and ethical behavior. You could counter his philanthropy and good intentions with loose business practices."

"She's got a point, Carl," green tie said. "Maybe we can still get some traction in a settlement. We'll agree to pull the story, but we'll get one of the stringers to recast the article with the facts we do have. We don't have to let Field's attorney's know. That's another story altogether. We can build off Briscoe's findings. It will be even bigger. Can't you see the marketing angle?"

Hollis couldn't read what Devi was thinking in his dark brown eyes.

He replied, "Just brilliant." The look he gave her was still unreadable.

After taking the next ten minutes to finalize a negotiating posture they wanted Mark and Hollis to pursue, Devi and friends left.

Mark packed up his papers. "You gave us enough space to breathe another day."

Hollis hurried to load her briefcase. "I was grasping. I've been trying to step back and look at Cathy's papers from all angles. Besides, with my back to the wall, it's amazing how creative I can get."

"I'll drink to that. Want to join me?"

"No thanks. I want to pick up those pictures from Joe, the photographer. We may not need them but he went to a lot of effort to make me copies."

Hollis finished putting papers in her briefcase. She felt on edge but couldn't exactly pinpoint the cause. Cathy's findings, the nonprofits and the meeting with Devi …. Things just didn't feel right.

"Hollis, for God's sake, can't you ever let go and have fun?"

She looked affronted. "I do have fun. I had a date last Saturday."

"With a guy?"

She put her hands on her hips. "Yeah, with a guy."

"That's great news. Now I can get Rena off my back. She's been at me to introduce you to a nice guy."

"Tell her thank you, but I found my own guy."

"Really?" Mark finished packing. "Do I know him?"

Hollis steeled herself. "It's John Faber."

Mark sat back down. "Detective John Faber? John Faber who almost sent you back to prison? That John Faber?"

"Very funny. He's actually kind of nice." Hollis turned away from Mark's astonished look. "Come on, let's go home. We made it through this wretched day."

CHAPTER TWENTY

———— ❧ ————

HOLLIS GLANCED AROUND the half-empty café and took a long sip of ice tea. The real crowd wouldn't show until closer to noon. She looked at her watch again. Kelly was late.

After yesterday's debacle with Eric Ferris, she was feeling deflated. On top of that, the less than stellar presentation for Devi made her want to crawl under the cover and hide. What had Cathy been thinking? Why would she have produced an article with such damning consequences and so little evidence—so much non-evidence? Hollis had always relied on her ability to see into people. Now, looking over Cathy's work, she wasn't too sure. Had she been wrong about Kelly, too?

She looked up as a gust of air preceded a young woman through the door. Hollis waved her over.

"Sorry I'm late. The hospital called me back last night. Granddad had a small heart attack. He—"

"Oh my God! I'm so sorry, Kelly." Hollis reached across the table and put a hand on her arm. She didn't bother to hide the misery she was feeling. She took notice of Kelly's own flushed cheeks against pale skin. "Is he going to be okay?"

Kelly closed her eyes and nodded. When she opened them, they glistened with tears. She remained silent.

Hollis stiffened. "I feel terrible. It was me. I brought it on. It was the letter from Margaret that upset him." Hollis didn't wait for a courtesy denial from Kelly. She knew she was right.

"It could have been worse. The doctor said it was a warning." Kelly wiped at her eyes. "Hollis, I had a chance to talk to Granddad this morning. He said he wanted me to know that … that he loved me … and that he was wrong."

"Wrong about what?"

"He wouldn't say. But I'm pretty sure it's because he lied." She shook her head. "He just kept repeating that he was wrong. Then, he fell back to sleep."

Hollis said nothing.

The waitress came to take their order. Kelly ordered soup, Hollis a tuna sandwich.

"All these years, I thought my grandfather lost his freedom because of Margaret Koch." Kelly stared out the window.

Touched by the young woman's pain, Hollis avoided her eyes by staring down at her own hands.

Kelly continued, "I hated her."

Hollis sighed and nodded in understanding. "Margaret thought that your grandfather was Charles' killer. She thought he and his brother had gotten into a fight over her."

Kelly continued to stare out the window, dabbing at her eyes with her now wilted tissue.

Hollis spoke softly. "Margaret Koch got her husband to use his influence to have your grandfather's case reviewed by the governor. They somehow got the evidence that Charles had set up his own suicide to look like murder."

Kelly held Hollis' gaze. "But why didn't Granddad tell my mom or *me*? Why continue the lie about Margaret? Do you think that's what he meant when he said he was wrong?"

"I don't know." Hollis paused, reflecting on the power of hate and love lost. "Eric Ferris confessed to a crime and went to prison because he thought he was saving the woman he loved. Except the woman he loved gave her love to another man. He has carried this … this pain until it physically broke his heart."

"But she got him out of prison."

"And he despised her for it."

Kelly shook her head. "I don't understand."

"He had grown comfortable with drinking out of the glass of bitterness. He didn't want to owe her. He didn't want her to give him a reason to forgive her."

"Do you think she knew?"

Hollis shrugged. "I don't know, but I have a feeling he didn't have to worry. She never forgave herself."

Kelly dug in her purse and pulled out a letter. She handed it Hollis. "I'm keeping my word. I've been holding on to this for myself. I wasn't ever going to show it to you. I wanted to tease you with it so I could see the letters you had." She wiped at her eyes. "But you showed me the truth about my family and that means a lot …. It means everything. " She hesitated. "It has sentimental value for me. It's from my grandmother to my Mom. Gram was dying but we didn't know it."

Hollis gave a rueful little laugh. "You know it's funny; I got wrapped up in Margaret's history. It was like I was watching a life unfold, a melodramatic life. The letters piqued my curiosity, and they were addictive. A couple of days ago I wanted nothing more than to read your letter. Even though it wasn't from Margaret, it would add to her story." Hollis still didn't reach for the letter. "Now it's my closure."

Kelly made a sympathetic face. Hollis took the sheet out of the envelope. It was written in 1995. The handwriting was effusive and thick.

Dear Mary,

Your father and I so enjoyed your visit last Sunday. Kelly is growing up so fast. She reminds me of you when you were two. I love you both so much.

You keep dodging me, but I want to write of things that speak of endings and beginnings. I want to say them in writing because after I'm gone, I know you will read my words over and over. If I spoke them, you would rely on your own interpretation and memory. My words would blur.

When I was a young girl, my cousin Margaret and I would play in the park in Rowan and imagine what it would be like to travel the world and eat exotic food. She was like a sister to me. Her mother, Aunt Louise, was stern and always had a frown, but my mother, your grandmother, would let us dress up and join us in our fantasies. Later, all our other cousins died of the flu, so when your grandmother died, the only family I had left was Margaret.

We were as close as could be. But that all changed.

Margaret hated Rowan. She was far more adventurous than I. I envied her bravery. She got married, but her first husband died young. Then she met your father's brother, George. They were in love. It was a shame you never met your uncle. He was a good man.

I never told you this, but your father was married to Margaret for several years. I guess for a lot of reasons probably related to guilt, Margaret and your father drifted apart. Anyway, they divorced, and over time your father and I found each other and fell in love. We've been in love for a very long time.

I find it so easy to write about the past as if it were all yesterday.

I know you and Cliff are going through a difficult time. Eric and I had our difficult times too. But if you can just wait for the sun to come up again, you'd be surprised how different things will look to you.

I'm dying, my darling daughter. The doctors say there's no more that they can do. My time on this earth is very nearly over. The cancer has won this battle, but it has not defeated me, because I had you and now there's Kelly. Anything I have of worth I want you and Kelly to have.

Take care of your father for me. He can be difficult and I know you two bump heads now and then. But you are so much alike and he loves you dearly. Keep close to Cliff. He's a good man. Let your love be bigger than both of you.

Kiss Kelly for me and give her my opal ring when the time is right.

With love and respect,
Your Mother

Kelly took out another tissue. "When the topic of my grandfather's prison term would come up, I could hear him telling my mother Lisbeth: 'let it go, move on,' but she never could. How could she? Grandfather never let it go. And then my mom died when I was twelve."

"What did you just say?" Hollis slid her plate aside.

"That my mother died when I was twelve?"

"No, what was your mother's name?"

"Lisbeth Mary. It's an old-fashioned family name. She was named after her mother. She was Lisbeth, too."

Hollis got out her notepad, flipped through the pages, and found what she was looking for.

"Kelly, your grandmother …. Do you have any other letters? Do you have any of her papers? Maybe a birth certificate?"

Kelly moved forward in her chair. "I can look through Mom's things. They're back at the house. Why? Why do you want to know about my grandmother?"

Hollis didn't answer at first. The links were coming together and pieces were lining up.

"Can I see that letter again?"

"Sure." She slid it over. "Is there something wrong? My dad died a few years after my mom. This was the ring Gram wrote about in the letter." She held out her right hand and the dainty opalescent stone was set off with a circle of diamond chips. "Hollis, what is it? What are you thinking?"

"Kelly, I didn't show you the letter Margaret received from Lisbeth. I didn't think to, since it didn't have anything to do with your grandfather. I don't know you that well, but I think you're going to need an attorney." Hollis picked up her pad of paper, scribbled a note, and put her purse strap on her shoulder.

"Wait. Tell me what's going on." Kelly started to gather her things together. "I just wanted proof that Granddad was innocent." She took a breath. "When Mom and Dad died, it was just Granddad and me. I was a just a kid but he did the best he could. I thought it was prison that had done him in, but now I know it was Margaret Koch."

"Kelly, I'd hold off with that thought. If I'm right, you're Margaret's heir."

"What?"

"I can't go into it. Just show an attorney the letter and explain what's happened."

"Can't I use your firm? My husband and I don't have a lot of money for a lawyer."

Hollis started putting her things back in her purse. "No, you can't use our offices. It would be a conflict of interest for us to represent you. But, I can give you a list of good law firms."

"Are you sure I'm related, Hollis?"

"Lisbeth and Margaret were first cousins. Your mom was her second cousin, and that makes you her third cousin."

"Third cousins can inherit?"

"Under the right circumstances, yes." Hollis stopped gathering her things. "I'll let my boss know that you're a possible Koch heir. There's a hearing scheduled but we'll ask for a continuance. We'll wait to hear from your attorney. There will be a couple of court visits and some papers for you to sign, but in a few months you should receive an estate check."

Kelly smiled broadly. "Wow, that's great. It will come in handy with Grandfather's medical expenses."

Hollis frowned. "I don't think I would let your grandfather know where the money came from."

Kelly looked blank at first; then the meaning of Hollis's words sank in. "You're right. He would never accept money from Margaret. He was so angry and hurt by her. I'm just realizing that for years he made up a story with Margaret as his scapegoat to hide the truth he couldn't live with. He didn't even tell my grandmother. You're right; it's probably guilt that has made him ill."

Hollis fell silent.

"Now, what's the matter?"

Hollis' mind was going over Cathy's cache of materials. Articles and papers that didn't make much sense, especially when she had been trying to force-fit them into the story, like square pegs into round holes.

She had it all wrong.

"Hollis?"

"What?" Hollis turned to face Kelly, who was staring at her, head cocked. "Sorry. I'll talk to my boss about you. In a week or so we'll let you know when you can come in with your attorney to go over paperwork. I'll call you in a few days, but I've got to go … now."

She gathered her jacket, patting her pocket for her car keys.

"Okay, I'll wait to hear from you." Kelly reached for her purse and walked with Hollis to the door. "Did I say something wrong?"

Hollis stopped. "No, everything is good. I just had a thought about which is worse, living with a lie, or knowingly selling it to others."

"You're not talking about Granddad anymore, are you?"

"No, I'm not."

THE HANDLE ON Hollis' car door was sticky.

This posed a troublesome question. Why hadn't it been sticky this morning when she parked in the outdoor lot after her meeting with Kelly? It was a little after dusk, and even though most of Triple D's employees had gone home, there were still a fair number of cars around. She felt a slight unease. She took a Kleenex out of her purse and wiped down the handle. She smelled the tissue. It smelled like glue.

A sickening feeling went through her as she recalled a TV show about vandals who use glue to disable the locks on a car. *Not now*, she thought. But the door opened easily and she got in behind the wheel. She was about to pull out of the space when she glanced in her side view mirror and saw a pair of eyes staring at her.

"Whaaaaaaaaaaa," she screeched.

"Hollis, I'm sorry. I didn't mean to scare you," Vince called out.

She turned off the car and rolled down the window. "Are you out of your mind? I could've had a heart attack. Where were you hiding?"

He looked desolate. "I wasn't hiding. I was sitting on the curb. You just didn't see me." He zipped up his jacket. "Besides, you left your car door unlocked, so I locked it."

"With what? Did you have peanut butter and jelly on your hands? You don't know why I left it unlocked. Maybe I wanted it unlocked. Haven't you ever heard that if you don't own it, it's not yours?" She knew she was overreacting, but her heart still pounded rapidly as her adrenaline made the slow path back to its gland.

"I just wanted to thank you for being so nice to me, but I guess I fell asleep waiting for you."

"You guess?" Hollis was having little success calming her voice. She took a deep breath. "Look, you have got to stop following me. What do you do all day?"

He shrugged. "I mostly help out at the food kitchen, make deliveries for the shut-ins, and clean-up at Fresh Start at night."

"Is your mom still in rehab?"

Vince said nothing but kept his eyes downcast.

"Look, I'm really tired, and I'm really sorry about your Mom. But you're not going to be my shadow. You're starting to creep me out." She turned back to look out over her steering wheel.

He nodded. "I get it. I'll leave you alone." He backed away.

Hollis sighed in exasperation. "Don't you have any friends?"

He paused and shook his head. "You're my only friend."

His words hit her with a bolt of recognition.

She stared at him, wondering where the air in her lungs had gone. She blinked her eyes, which started to blur, and then opened them to see Vince, head down, hands deep in the pockets of his hoodie, make a right turn down the street.

Hollis took a deep sigh, started the car, and pulled up alongside him. "Okay, okay, get in."

Vince started to open the car door but saw her look at his hands. He gave them a brisk rub on his jeans.

Hollis rolled her eyes and pulled back into traffic.

"Are you hungry?"

Vince shook his head. "You think I just see you 'cause I want a handout?"

"Well …." The thought had crossed her mind.

Vince's face turned red.

Hollis felt shame. "No, no I don't think that. I just want you to be … safe." She spoke hurriedly. "I want you … I want you to have a real life."

He turned away from her and looked out the window.

"Vince?"

He turned to her, his hands making fists. "Let me help you. I can do a lot of stuff. Let me help you find the guy who killed your friend."

It was Hollis' turn to shake her head.

"I have a job that pays me to work on Cathy's murder. I don't have the money to pay you to work on it too."

He looked at her. "Not for money. I just want … I just want …." His voice drifted.

She nodded. Her voice caught. "I … I understand."

There was silence between them as she made her way through the streets and parked in front of a community park down a short ways from Triple D.

"Tell you what. You pass the GED, and I'll come up with something."

"Aw, Hollis, I don't need no GED."

She shook her head. "It's 'I don't need a GED'. If you had a GED you'd know that. I told you before that I'd help you. If you want to work with me then you have to have a high school diploma."

Scowling, he jerked on the door handle, got out and slammed the door. Hollis called after him, but he didn't look back as he blended into the crowd of pedestrians.

THE NEXT DAY, Hollis and Stephanie met for lunch in a little deli not far from the police station. Clearly, it was also the favored eatery of the forensic unit, because Stephanie got several nods of hello.

Stephanie leaned over to reach for the Italian dressing. "Well I'd say he was pretty ungrateful. I can't believe you're still dealing with that kid."

Hollis shook her head. "I'm not dealing with him. He's adopted me. Besides—"

"Besides, what? There are no besides."

"I don't think he has anywhere to go. I think he's alone. I'm even beginning to wonder about the existence of his mother."

Stephanie took another bite of salad. "Great, now you're going to start bringing home stray puppies."

"Will you stop saying that? He's not a stray puppy and I'm not bringing him home." She took a sip of water. "Let's talk about something else. How's work?"

"We're already down five people. Now they want to lay off one more, just as we're taking on more lab work from the Sheriff's Department." Stephanie's brow furrowed. "So, what's with Cathy's case? You any closer to clearing her?"

Hollis shrugged. "Mark and I are fairly confident about beating the suit. It's not our call, but I would hate to go with a settlement. I can't help thinking I'm missing something."

"You said 'fairly' confident. I know you. That's not good."

"No, that's not good, but it's where we are."

"What's the problem?"

"Cathy was a pro and had anyone asked me before she died, I would have bet my life that anything she wrote was not only verifiable and credible, but defensible. But I read her article. It was atypically harsh. She was writing a three-part series, so why did she come out with all barrels blasting in the first one?" Hollis played with her napkin, ripping the edges into thin shreds. "That's when I stepped back and said, 'What's wrong with this picture?' I can't go into it, but Stephanie, I think Cathy was onto something."

"What are you saying?"

"I don't know what I'm saying. Just that the article didn't sound like her." Hollis sighed. "Cathy hid papers that appear to have nothing to do with Fields. All my follow-up contacts leave me certain that something was going on. But I don't know if these papers had to do with Fields or if she was working on her next article."

"Maybe both. You know, when you look back at what happened, no matter what happens, you did the best you could." Stephanie looked at her watch. "I've got to get back, but I need a big favor."

"What?"

"Can I borrow your car on Sunday to go to a conference on Monday?"

"Again?"

"What? Wasn't the last time over six months ago?"

"Yeah, and I ended up with an empty gas tank, a parking ticket, an angry neighbor who swears it was me who ran across a corner of his manicured lawn with my tire, and you and I don't even look alike."

"That's not fair."

"I know, to me," said Hollis. "When are you going to get your own car?"

"I'm going to." Stephanie finished chewing on an olive. "I just can't make up my mind. I still can't get over losing Cooper."

Hollis almost choked on her bread crust. Stephanie had named her car Cooper because it was a bright red and white 2006 Mini Cooper.

"You didn't lose Cooper, you totaled it." Hollis laughed. "And that was almost two years ago. All your friends know you rotate us for our cars."

Stephanie looked sheepish. "So, can I?"

"Yes, yes. I'll get Mark to take me to the office on Monday," Hollis said. "I'll drop the car off Sunday at your house and you can come with me to the bookstore."

"Thanks, sweetie. I won't point out that you seem to be avoiding letting me inside your house. Maybe one day I'll be worthy." Stephanie got up and kissed her on the cheek.

Hollis started for the door. "And, Steph, I need it back no later than Tuesday."

Stephanie waved her away. "I'll make it Monday."

CHAPTER TWENTY-ONE

⟨⟨⟨⟩⟩⟩

O N MONDAY, SETTLED in her desk chair, Hollis read *Transformation*'s draft rewrite of Cathy's article. She had to admit the reporter had done a good job. She made a couple comments at the end of the text and then scanned the copy to Mark so he could add his review. Fields' attorneys were still trying to have all their legal fees paid by *Transformation* magazine; Carl Devi was just as focused on making sure they didn't get a dime.

Mark suggested that Devi let them see the rewrite after incorporating their comments, but Devi nixed the idea. He was not going to be censored. It set a bad precedent. Hollis shrugged. It was his money.

She was eating lunch in her office, because Stephanie had her car and wouldn't be bringing it back until the end of the day.

Her ring tone went off and she rustled through her purse for her phone.

"Hungry?"

"Why, Mr. Faber, isn't it a little late for lunch?"

"Well, not lunch. I've got to follow up on a case; I was hoping you'd be interested in an early dinner."

"I'm sure I'll be hungry then, too." Hollis smiled. "Actually, that will work out better. A friend borrowed my car, and now she won't have to pick me up. Can you meet me at my office?"

"See you out front."

The smile remained on her face even after she hung up, and it was still there after George appeared in her doorway.

"Boy, something has lit up your face."

Hollis started and the smile vanished. "George, you scared me. What's the matter?"

"Not a thing. I was able to get a continuance in the Koch probate hearing. I explained to the judge the last minute confirmation we received. There wasn't a problem. But don't think I've forgotten you didn't take my calls the other day." He took a seat across from her. "You're only off the hook because you found an heir to the Koch estate. But don't ignore my calls again. Still, I want to reiterate: it was real good work, Hollis. I'll have to let Pierson know that his report was faulty."

She winced. She was glad she didn't have to be the one to tell him.

"Thanks, George," she said, not bothering to suppress her smile of gratification. "I'll redo all the filing papers and get them over to the court early next week."

"I assume I no longer have to worry about further delays?" he said. "I got your email about the *Transformation* matter. Once that gets wrapped up, I could use you on a complicated estate trust. You need to contact the parties and make arrangements for a meeting in two weeks."

"Got it." She scribbled down notes. "George, I apologize for the time it's taken for the *Transformation* lawsuit. It's about over. Mark thinks that *Transformation* is going to pull the defense plug."

He nodded. "I'm sorry. I know you wanted to clear her reputation. From what you've told me, the proof just wasn't there." He looked at his watch and stood. "I've got a client meeting to go to. Let's you and I get together Thursday at three so I can turn this new matter over to you."

As soon as he left, Hollis called Stephanie and told her to drop the car off in the driveway. She could give back the spare key at lunch the next week.

Stephanie's smile was evident in her voice. "Dinner again, huh? Sounds like Mr. Faber wants to be a contender."

"Oh, no. I've been down that road before."

"I wouldn't paint everyone like your ex."

Hollis didn't answer.

It was easier said than done.

THIS TIME HOLLIS felt more confident and composed on their date. John picked a local eatery that specialized in home-cooked Chinese food, and they both realized they preferred to use chopsticks. The evening had gone smoothly. They talked about everything but their work, and they slowly relaxed into a comfortable silence.

Hollis leaned back. "I meant to ask you this before. Why did you go into law enforcement?"

"Two reasons." John pursed his lips. "It was television. I loved the action. And my uncle, who I admired a lot as a kid, gave me the idea that helping people by putting away bad guys would be rewarding." He picked up a shrimp, bit half then popped the rest of it in his mouth.

"Your uncle was a police officer?"

"Yep, he served on the force for twenty-three years before he was shot and killed by one of those bad guys trying to rob a store."

"I'm sorry."

Neither of them said anything for a moment. It wasn't an awkward silence—more a contemplative one.

John took a swallow of Tsingtao. "I know we talked about this last time, but why probate? I mean really, why?"

"I've settled on probate law because it's what I know best."

She dabbed her mouth with the napkin. "For me, it's a good balance between people and problem-solving."

"Aren't your clients dead?"

Hollis laughed. "I never saw it that way, but I see your point. No, my clients are also the families left behind and the estate itself. They all need closure; I bring closure. It's like I'm the last chance to get it right."

"That makes a lot of sense." He paused. She could tell he wanted to say or ask her something.

"What?"

He looked her in the eyes. "Why did you do it?"

The question took her off guard. Then a combination of anger and shame shot through her.

"You mean the insurance fraud?" She bristled. "I thought we had this conversation a long time ago."

John nodded. "I know I risk you leaving me at this table. But ordinarily I don't get to ask that question. I don't even want to know the answer because I usually don't care, but I care about you."

She fingered the porcelain tea cup. She wasn't sure she was going to answer until she heard herself respond. "I turned my back on me. I put someone else's idea of who I was in place of my own moral compass. I just plain screwed up."

He stared into his glass and said nothing.

She didn't realize how much she wanted to have him answer, until he didn't.

The waitress put the bill between them.

Hollis snatched it.

John protested. "No. This was my invite. Why are you doing this?"

"Because I need to. Please take me home."

He finally convinced her to give him the check, but the atmosphere between them was strained and thick. Although he made several attempts to get her to talk while they drove, she refused.

"Hollis, I didn't mean to upset you. Can't you give me another chance?"

He reached for her hand and she let him take it.

"I guess I just plain screwed up."

A slow sad smile crept across her face. She nodded.

"We're okay?" he asked.

She nodded again.

THEY PULLED ALONG the curb in front of her condo. Her car was in the driveway where Stephanie left it. Hollis frowned. Stephanie had left the car's interior light on and the door cracked.

"Well, it looks like we got here in time before the battery went dead," John said, getting out of the car.

Hollis frowned. "She's not usually this scatter-brained."

She walked up the driveway to the front steps and screamed. John scrambled after and looked around the car to where Stephanie lay across the bottom two entry steps. He dialed 911.

Hollis bent down. The scent of honeysuckle drifted into the night air.

"Wait," John said.

She pulled back.

He put his fingers against Stephanie's neck and spoke with the dispatcher on the phone.

"She's still alive. Don't touch her; the medics are on their way."

She bent over her friend. "Oh, my God, Stephanie, can you hear me?"

No response, but sirens were already approaching. She looked at John.

"What do you think happened?"

"Judging by what I can see of her injury and the pool of blood under her, I'd say she was shot."

Shot.

An ambulance pulled up behind the car and two EMTs

jumped out. John pulled Hollis aside, but she moved forward when she heard Stephanie groan.

"Steph, can you hear me? I'm here."

They were still lifting her into the ambulance when a patrol car and unmarked vehicle pulled up. John went over to speak to a tall, thin man, who Hollis gathered from experience must be the lead detective.

Hollis went up to the EMT as he was pushing the ambulance doors shut.

"Can I ride with her?"

"Are you family?"

"No, she doesn't have any family. I'm her best friend."

"I'm sorry, we can't let you ride inside, but we're taking her to St. Rose."

She stepped back onto the sidewalk.

John came to her side. "Hollis, can you take a minute and talk to Detective Perry about what Stephanie was doing here with your car?"

The older man standing behind John came forward.

"Detective Jake Perry, Ms. Morgan. Sorry about your friend, but I have a couple of questions."

"I … I can't. I've got to go to the hospital."

"Yes, ma'am. We can get this over with real quick."

Ignoring him, Hollis walked away to get her purse out of John's car. She looked back and realized she didn't have a car. John came up to her and leaned down. "Let me drive you."

She looked into his eyes. "Thank you, can we go now?"

He pulled her to the side. "Answer Perry's questions. I'll tell him to limit it to two; he can save the rest for tomorrow. Besides, you won't be able to see Stephanie until they examine and work on her. You can help her best by helping Perry find who did this thing."

It was dark now. The neighbors who had been standing on their front lawns had started to go inside. Hollis noticed the officers knocking on her neighbors' doors. She sighed and walked around to the side pathway leading to the rear yard. The next homeowners' meeting would likely include a petition to make her explain the steady flow of police activity at her home.

John motioned to Perry to follow. The three stood on a little patio at the side of the building.

"I'd invite you to sit but I need to get to the hospital," Hollis said.

It was clear to her that this setup was less than desirable for both of them, but she didn't care.

Perry pulled out a pad and pen. "Ms. Morgan can you explain why Stephanie Ross was driving your car?"

Hollis explained with what she hoped resembled patience.

John gave her an encouraging nod.

"Okay, did you tell anyone she was going to be here?"

Hollis hesitated. "No, no one knew. Did anyone see anything or hear the shots?"

"We're checking now, ma'am. But maybe you could tell me—"

John put his hand on Perry's shoulder. "Jake, you had your two questions. I need to get Hollis to the hospital. I'm sure she would be willing to meet you at the station tomorrow to answer the rest."

Hollis' heart did a grateful leap. She gave him a look of immense gratitude.

Perry shrugged. "Okay, be in my office at nine a.m." He gave Faber a curious look and walked back to his car.

John turned her to face him. "You can't stay here tonight. Is there a friend you can visit?"

She shook her head.

Cathy was dead, and now Stephanie had been attacked.

"Well, let's get to the hospital and we'll talk about lodging later."

ST. ROSE HOSPITAL was much nicer than Hollis expected.

Instead of plastic chairs, the waiting room was furnished with sofas and loveseats. But it was still a waiting room, and so she waited. John left her while he went to check on Stephanie's injuries.

It had been almost four hours since they found her.

Hollis was grateful she didn't have to share the waiting room with strangers. She was deep in thought when she felt a touch on her shoulder. It was John.

"She's still in surgery. She was shot in the chest. The bullet hit an artery but it was a through and through. Sorry, a through-and-through is—"

"A through-and-through means it was a clean shot," Hollis said. "It went in and came out. I've been to the big house, remember?"

He gave her a small smile. "Anyway, she's holding her own. The doctor knows you're waiting, and he'll come out when they're finished."

She grimaced. "Why would someone shoot Stephanie? Could she have interrupted another burglary?"

John hesitated. "She was shot before she entered the house. I don't think it was your friend they were waiting for."

Hollis met his gaze. "They were waiting for me? But why?"

He sat beside her. "When your place was broken into, drawers and cabinets were tossed but nothing was taken, not even your laptop. But they took the time to batter it." He picked up her hand. "Someone thinks you have information that could hurt them."

She just stared at him, but slowly the answer dawned on her. "It must be Fields."

"Tell me about this article Cathy Briscoe wrote."

Hollis went through the story line. "I gave up finding any real dirt. We're in the process of making a settlement."

"From what I read about him, the guy seemed to be one of the good ones, but if he's afraid that you could ruin his reputation, he might be willing to kill to stop you."

They both looked up as the doctor walked into the room with a clipboard.

"You're here for Ms. Ross? She's going to be fine. It was more messy than serious. The bullet missed the aorta. She's conscious but she'll be off to sleep soon. You can see her later this morning."

"Oh, thank you, Doctor." Hollis got up and shook the doctor's hand. "How long will she have to be here?"

"Just a few days. Does she have someone at home? She'll need help with meals and return visits."

Hollis brightened. "I can stay with her. I'll bring her back for her checkups."

JOHN HELD THE door open for Hollis as she climbed into the car. "Uh, don't take what I'm about to say wrong, but I have a sofa you can sleep on. You can get a couple of hours sleep, then I'll take you home. They won't have to impound your car and you can get into your condo."

She could feel her heart start pound with a staccato rhythm. "I … I …."

"I know what you're going to say. I have an early shift today, so I won't be home except to shower and leave again. I plan on checking with Perry to see if the neighbors saw anything. Then I'll come for you and take you to back to the station to have your talk with Perry."

"No, I didn't mean that … I …. Yes, thank you."

She wondered if she would ever stop sounding like a babbling idiot.

They drove in silence. He lived in a small ranch style house on the edge of Castro Valley. Inside it was clean and had a certain charm if you had a preference for 1970s jazz musician art. Too tired to peruse the bookshelves stuffed with books, Hollis gratefully took the blankets and pillow to make her bed. She set the alarm on her cellphone and was asleep in minutes.

***

THE NEXT MORNING they stopped outside the large police station anteroom that led to security. Hollis called George to brief him about the shooting and to tell him that she would be late to the office. He told her to just take the day off. She was appreciative but she wanted to keep her mind busy with filing the new Koch papers.

She rubbed her arms. She should have brought a jacket. It was warm outside, but the station air conditioning was blowing an arctic cold. John had dropped her off a few minutes before nine. He offered to come with her to the meeting with Perry, but Hollis felt he had already done more than enough to help.

"I've got to wrap up some things, but I'll be checking periodically to see if you're done." He accompanied her down the corridor. "I don't think I ever got the chance to tell you what a great time I had yesterday."

"Me, too." Hollis smiled. "My life really is not this complicated. It seems like you're always catching me in the midst of a crisis. But I'm glad you were there to help me with Stephanie."

"Uh, I can think of some pretty good moments."

"Good, then maybe you'll let me cook a meal for you."

"Say no more. You got a deal." He bent over and kissed her on her forehead; then he looked into her eyes and kissed her lightly on the lips. "I'll be back when I think you've finished."

Faber had no sooner turned the corner when Perry entered the lobby. "Ms. Morgan, I appreciate you coming here. Take the hallway in front of you. My office is the second door on the right."

He walked behind her.

She took the only chair in front of the desk.

"I understand that Stephanie is on her way back to health. That's real good news. She's excellent at what she does. We miss her already."

Hollis gave him a tentative smile. "Yes, she's doing well. I spoke briefly with the doctor early this morning. She may only have to stay a few more days."

"Good, good." He scratched his neck. "So, how do you know her?"

The next few minutes Perry took notes and asked follow-up questions. He didn't look up.

"So, this wasn't the first time she'd been to your home?"

"No."

"Detective Faber thinks the shooter was after you. What do you think?"

Hollis paused. "I think so, too."

He looked up. "I did a little research. You served time."

She could feel the heat climbing up her neck. "If you know that, you know that I received a pardon."

"Yeah, a pardon." Perry pulled on his ear. "So, why would someone want to kill you?"

"I think this is linked to the Briscoe murder, Detective. I must have—or the killer thinks I have—some information that could incriminate him, or her."

"What kind of information?"

"That's just it. I don't know."

He rubbed his hands over his head.

"You said you got a file from Briscoe—articles, sketches but nothing significant? Nothing that made any sense to you?"

"That's right. I gave it all to Detective Cavanaugh."

"Well, Cavanaugh is close to retirement," Perry snorted. "He's not happy about this attempted murder at all. He's likely to be assigned to this case because he has the two murders. I'll get with him later. Now … you said you found some papers later in this hiding place in Briscoe's apartment and gave those to Cavanaugh too?"

"Yes." She didn't think he needed to know she was doing her own investigation. "Did either of you speak with her boyfriend?"

He rubbed his hand over his balding head. "Ah, yeah, the so-called boyfriend you ran into."

"Right, Michael Carver."

"Funny thing," he said, rubbing his palms together. "We can't find any evidence of a Michael Carver. He's disappeared."

CHAPTER TWENTY-TWO

———

HOLLIS STARED OUT the window waiting for Stephanie to return from X-ray. The view from the hospital room gave an uninterrupted vista of the faux brick façade of the apartment building across the way.

She mulled over her conversation with Perry. He went on to tell her he was trying to decipher the true identity of Cathy's boyfriend, Michael Carver, if he really was her boyfriend. Hollis had tried to contact Carver using the number he gave her, but it led nowhere. Her fears were confirmed when Cathy's true beau contacted Evelyn Briscoe, wanting to know about the funeral arrangements. Evelyn had given him Hollis' number. Evidently, he was still within the ninety-day probation period.

Hollis didn't doubt that Carver was sent to make sure no other evidence of Cathy's notes remained.

Her reverie was broken by approaching voices.

"If you didn't want me to borrow your car, all you had to do was tell me." Stephanie walked gingerly on the arm of a nurse while hanging onto a mobile rail holding two bags of liquid and long tubes that ended in her arm and the back of her hand. Her voice was weak.

Seeing her friend, tears streamed down Hollis' face.

"Oh Steph, I am so glad to see you, and I am so sorry this happened."

She helped the nurse get Stephanie re-settled into the bed.

"What are friends for?" Though she was trying to put up a good front, Hollis could see the strain of pain in her expression.

"Do you remember what happened?"

"Not really. I was about to put your newspaper on the landing. Then, someone—a man—came from the side of the house holding out a .38 caliber. I examine tens of those a week, but nothing brings it home like having it pointed at you. I don't remember hearing the shot or falling."

"It should be me in this bed, not you."

"I'd give you the chance, but like most things, it's non-transferrable. I'm going to be all right, just a little tired and sleepy."

"Are you in a lot of pain?"

Stephanie managed a little smile. "Nothing I can't handle. But there's something I want to ask you. The doctor told me that I could leave day after tomorrow, if I had someone to stay with me." She licked her cracked lips. "If—"

"Of course I'll stay with you." Hollis wiped her eyes with the back of her hand.

"Good," Stephanie's voice faded. "My keys are in my purse. The nurse will get it for you. I don't want you staying in your condo. Promise me, Hollis." She laid her head back down.

Hollis' eyes filled again with tears and she nodded. "I promise."

Stephanie closed her eyes.

"She'll sleep." The nurse motioned for Hollis to meet her in the hallway. "She's a little too optimistic. It may be more than a couple of days before her release. But she's making good progress." She handed her a plastic bag. "Here are her keys and her wallet. It's best you take them with you."

With one last look to Stephanie, she hurried out of the hospital.

***

HOLLIS ENTERED HER condo tentatively, as if she were a visitor to her own home, and quickly and quietly went through all the rooms. She just as hurriedly packed a bag of clothes. Whoever shot Stephanie apparently hadn't tried or at least hadn't succeeded in entering the house. The thought flashed through her mind that Michael Carver or whatever his name was could easily have been the culprit. She took a last look around her living room and locked the door on her way out.

At Stephanie's, she emptied her suitcase onto the bed in the spare bedroom and looked around. It was a good-sized room with the basics for a guest who shouldn't count on staying too long. She placed her things in one of the two empty dresser drawers and hung up her clothes. She'd brought enough clothing to get her through several days.

She picked up Cathy's files and headed to Mark's office.

MARK, WHO HAD decided that coming into the office on Sunday would give him a head start on the week, listened in worried disbelief while Hollis recounted Friday night's events. "If you're frightened, I can get you a patrol officer drive-by."

"Already done." Hollis sat comfortably in the overstuffed chair in his office with a hot cup of tea. "I'm not frightened, I'm angry. The only good thing about these horrible events is that they're keeping me from obsessing over waiting for my bar scores. I feel like I'm being jerked around because I'm missing the whole point." She briefly closed her eyes and gave him a grim smile. "Besides, I told Stephanie I would stay with her."

"Good," Mark said. "I don't like the idea of you staying by yourself. First the burglary, now this. Things are escalating. Also, there's been a major snag. The settlement is off. Field's attorneys weren't impressed with the article rewrite. They came back with 'too little, too late.' But I think they have really overplayed their hand." Mark tossed papers from his briefcase onto the desk.

"I notice they didn't withdraw until after *Transformation* rewrote the article."

Mark nodded. "Yeah, so much for fair play. Devi is not pleased. Still, it's not clear that Cathy was really on to something. On the other hand, there's *something* going on, because people are getting killed and hurt. How does the attack on your friend Stephanie fit in? Scratch that, the attack was meant for you."

"There's more." Hollis took a sip of her tea. "Cathy's 'boyfriend' has dropped out of sight."

"What!"

Hollis nodded. "It's clear now. Carver fooled me into thinking he was her boyfriend. He must be on Fields' payroll. I think he was trying to wrap up any loose ends. Then last night he tried to wrap up one big loose end—me."

"Using a gun. That's the move of a desperate man." Mark frowned.

Hollis mused, "Fields is in panic mode. I must be getting close, too. Unfortunately he's giving me more credit than is due. I don't know what I'm looking at."

"Step back and start over," Mark said. "What have you got?"

"I've been kicking around this theory. Cathy finds out information about how sloppily Fields runs his non-profits. She makes contacts from the charities themselves to the donors at the fundraisers that benefit them. If you follow the money, I think she found out something about a donor."

"Go on."

"She discovers it's not the charities that are the story, it's a donor. Someone with a secret he or she would do anything to protect, even kill."

WHEN CAVANAUGH SHOWED up at her office early Monday morning, she was prepared. He had called ahead to ask if he could "stop by."

"Boy, Ms. Morgan, you don't exactly live a boring life do you?"

"No, Detective, I don't. To save you time, I don't know who tried to kill Stephanie, or rather, kill me. I told Detective Perry everything I know."

They were sitting in the firm's conference room.

"Yeah, he's pretty sure you did."

"Have you found Michael Carver yet?" Hollis asked.

"No, and I can tell you when we do, he won't be using the name Michael Carver."

That's for sure.

"What happens now?"

"First, we are going to give you regular surveillance. I don't have the budget to give you twenty-four-hour protection, but we'll have units pass by your home on a regular basis. Let us know if you think anything looks or feels suspicious."

Hollis frowned. "I'm going to stay with Stephanie Ross while she's recuperating. But I don't think I like the idea of having the police follow me around."

"There have been two murders and an attempted murder. You don't have a lot of say in this. I don't like the idea of having another murder on my turf." He took his cellphone out of his pocket and glanced down. "Staying with Miss Ross is a good idea. She's a good employee. We all like her and she lives close to the station."

She nodded in acceptance. "What about Fields? Are you going to arrest him?"

"We questioned Mr. Fields, and he has an alibi." Cavanaugh was wearing his best poker face. "Let me worry about Fields. You take care of our girl."

STEPHANIE WAS PROPPED up in bed when Hollis visited the hospital on her lunch hour. The color had returned to her face, and her blond hair was gathered in a loose ponytail.

"Hey, thanks for coming by. Did you get all your stuff moved in?" When Stephanie tried to sit up in the bed, she gave a slight grimace of pain, followed by a rueful smile.

Hollis felt a new pang of guilt. "Yes, I did. And everything is all ready for your homecoming, day after tomorrow. Do you know what time they are going to set you free?"

The doctor entered. "You can pick her up at eleven a.m. We need her bed for sick people."

Stephanie smiled. "I agree."

This was not the doctor Hollis had met earlier. He was short and slim with the air of efficient competency. He extended his hand. "Hello, are you my surrogate?"

Hollis laughed. "Yes, I'm ready for my nursing orders."

He talked her through care instructions while checking Stephanie's wound. Hollis held her breath as the bandages came off, and although she wasn't squeamish, it still hurt her to see her friend deal with pain that should have been her own.

"Things look good. She's healing nicely. Bring her back Friday and I'll take another look."

He left with a wave of his clipboard.

Stephanie turned to Hollis, who stood at the side of her bed. "I know you're busy helping Mark with Cathy's case and assignments from your own job." She pointed to her bandages. "Are you sure you're going to be able to do this?" Her voice cracked. The examination must have tired her.

"Not even a concern," Hollis said. "I'm done with the Koch matter and I'm onto a new case George just gave me. *Transformation* is moving forward, and Cathy's case has been taken over by the police."

She knew Stephanie wouldn't pick up on her lie about *Transformation*, and after debating with herself, she decided not to mention the police surveillance until Stephanie was released. It also didn't seem a good time to tell her that she was still pursuing leads in Cathy's murder.

"You don't fool me." Stephanie said as if reading her mind. "Be careful, Hollis."

CHAPTER TWENTY-THREE

———∞———

"JOE, REMEMBER WHEN I asked you about this receipt for photos? You said some looked like morgue shots." Hollis pushed the receipt across the counter.

He picked it up. "Yeah, I remember."

"I don't suppose you have negatives?"

"Nah, it was digital. Cathy wanted fast and cheap. Unless I'm doing the shoot setup or it's an ongoing customer, I don't keep those things."

"Did any of the people look familiar?" She didn't want to lead him, but she had to know. "Did any look like Dorian Fields?"

"The philanthropist? Nah. Him I know. Every once in a while, Fields of Giving hires me to take photos of their special events."

Hollis perked up. "What was the last event you shot?"

Joe wrinkled his brow. "It was about three, no, maybe four months ago. It was a fundraiser over at the Camellia Mansion. I only charged them half my fee. I do that sometimes when I think it's for a good cause."

"Anything unusual happen?"

"Like what?"

"Did everything go as planned? Was the event a success?"

"Yeah, I guess so. I remember that guy who works for Fields really got on my nerves. He was always fussing around trying to tell me how to do my job."

"Who do you mean? Wade Bartlett?"

"Yeah, him. I was ready to punch him out. I finally ended up telling him off and to let me do what I know how to do."

"What happened?"

"He left me alone."

Hollis mentally racked her brain for the right question to ask that would push open the door to her nagging feeling that something wasn't right about Cathy's article.

"Joe, the photos you told me about earlier, do you remember where you shot them?"

"No, sorry I don't re—"

"Wait," Hollis broke in, "wouldn't you put it on your calendar or appointment book? Couldn't you check that?" Hollis' voice rose.

"Yeah, yeah, I can look there." Joe moved to his desk and picked up his calendar book. "Let me see, the first photos were taken right before Mardi Gras, because afterwards I went to New Orleans. Cathy insisted they had to get done before I left. So that's February ...."

He flipped pages and ran his finger down a column. Hollis practiced being patient.

"Here, here it is. Outside the Remington Building. I remember now."

Hollis was confused. "The Remington Building, that's the one in the financial district? Who did Cathy want you to shoot?"

Joe rubbed his chin. "You know it was strange. She just told me to be out front and out of sight. Starting at one-thirty, for the next thirty minutes I should take shots of whoever came out of the building. So I did. Less than ten minutes later these three guys come out talking and gesturing. I could tell there was some kind of argument. I used my zoom lens from across the street."

"Would you recognize them again?"

"Maybe … but maybe not."

Hollis could feel her frustration rising.

"You couldn't hear them at all?"

"Just a word here and there, nothing that made any sense. I took about a half dozen shots. But I remember Cathy was pleased when I turned them over."

Hollis made notes on a pad. "Was Wade Bartlett one of the men on the steps?"

"Nah, him I would recognize."

She put her pen and paper away. "Thanks, Joe. I have another favor to ask. If I came back with some photos, could you let me know if any of them are of those men who stood on the steps arguing that day?"

He shrugged. "I could try."

Hollis made her way back to the car, running through a checklist of suppositions. Something told her that Joe's photos held the key to Cathy's death. If Cathy's story about Fields wasn't what it appeared, there might be another story, a bigger story that Cathy somehow stumbled across. Suppose she had gotten close to the truth—only it was nothing she was ready to reveal that night she came to ask for Hollis' help.

Suppose there was another story—a deadly one.

CHAPTER TWENTY-FOUR

SITTING ACROSS THE polished walnut table in the *Transformation* conference room, Dorian Fields appeared more like a kindly grandfather than the plaintiff in a libel case. Wade Bartlett sat next to Dorian. He gave Hollis a nod and flashed a smile.

Hollis sat next to Mark, who was to Carl Devi's right. Polite small talk was quickly dispensed with.

Fields was the first to speak. "Let me understand, Mr. Devi, you're offering me a deal where if I accept your so-called offer, you'll pretend this whole thing was a … a mistake. Do I have that correct?" Fields' brow wrinkled and his white-gray eyebrows knitted together. "Clearly you have your own doubts. I read your retraction. It was well done and for once had kernels of truth."

Devi glared. "No, that's not correct. You accept this deal and what we publish will only focus on your poor business practices and questionable oversight." Devi practically sneered. "We haven't come out with our most damaging findings. You'll be an outcast. Your donors will dry up."

Hollis and Mark exchanged worried glances.

Fields was seemingly unfazed by Devi's comment. With a small smile, he flipped through the settlement agreement.

"Well, you know what?" Fields tossed the pages on the table, scattering them in front of Devi. "You can keep your agreement. I'll have no part of it. Fields of Giving does good works. It has real people who count on it every day to be there for them. You can come at me with your lies, but it will not stop me. You can't prove a thing. But rest assured, I will continue to take my case before a jury, and it will be you, Mr. Devi, who will regret this ill-conceived strategy."

Carl Devi stiffened, his face turning a deep red.

"Then I can promise—"

"Goodbye, Mr. Devi." Fields stood and started to move toward the door. "Mr. Haddan, Ms. Morgan."

He and Bartlett were gone.

Hollis and Mark said nothing.

Devi stayed silent until he reached and gathered up papers. "Well, Ms. Morgan, I guess we owe you for setting us up for this catastrophe. *Transformation* cannot fight this suit. We have little to no defense." He replaced the cap on his Montblanc pen. "You and Haddan have wasted thousands of dollars trying to defend your friend from shoddy reporting, and as a *coup de grace* you pawn off a preposterous idea that clearly will not pass muster."

Mark leaned over. "Wait a minute! That's uncalled for. We tried to help, and Hollis did a great job of follow-up and follow-through. Those centers on that list you gave us left us with a lot of questions. We had a fighting chance at a settlement. If your negotiating skills and bedside manner turned them off, it wasn't our fault."

Carl Devi stood. "Just send us your final billing. I see no need for us to meet again. We can take things from here."

Hollis sat, stone-faced. She gave a little shake of her head to Mark when he started to block Devi's exit.

She turned to him. "Mr. Devi, I am a professional and I

assure you that if there was any evidence to substantiate Cathy's story, we would have found it." She swallowed. "No one wanted to find that evidence more than I. You're right; Cathy was a friend. However, the fact is your legal team let that original article get published without hard proof."

"I'm not going to play the game of pointing fingers with you, Ms. Morgan," he snapped. "If you are as professional as you say, you'll accept responsibility and move on. We chew up lawyers like you. Oh, that's right, you're not a lawyer yet—or maybe just not a good one."

"Wait a minute," Mark objected. "I was the attorney in charge. I'm the one who should have known better. The responsibility is mine."

Carl Devi raised his eyebrows. "You're right. You're incompetent, too."

He slammed the door behind him.

Mark began, "Hollis—"

"Mark, don't. I'm a big girl. I'm not going to say he didn't get to me, but we let our feelings for a friend blind us to the facts." She sighed. "Cathy didn't have a story she could support with facts."

Mark was silent. They packed up the papers and files into their briefcases.

He turned to her. "We could both use a drink. Unless you agree to come to dinner with Rena and me, I'm going to keep on trying to cheer you up."

Hollis held up her hand to silence Mark and went over to the receptionist. "Excuse me, can I get a copy of *Transformation's* latest annual report?"

Anti-social Phil, who was madly texting, pointed over to a magazine stand in the corner. "You can find last year's over there. We won't have this year's for another month."

"Last year's is fine." Hollis smiled.

Mark came up to her. "What are you doing? Why do you want a *Transformation* magazine annual report?"

"I have a long-shot of a hunch."

"Let me know what it is before I have to bail you out of jail."
He shook his head. "Now, will you come to dinner with Rena
and me?"

"I've got to take care of Stephanie."

"I'm suggesting dinner, not a vacation."

Hollis gave him a small smile. "That sounds like something
I would say."

THEY HAD DECIDED on a mutually acceptable Italian
restaurant in North Beach, not far from downtown San
Francisco. Hollis tried to be companionable, but too much had
happened over the last few days and she knew her mood was
glum. Dinner was quiet and soon over.

Rena looked from Mark to Hollis and put her final forkful in
her mouth. "Okay, do you guys want to talk about this or are
we going to end the evening in the cellar?"

"Sorry," Hollis said. "I told Mark I wouldn't be good
company."

"Hey, I admit we've had a few tough weeks," Mark said, "but
we gave it our best shot."

"Let's change the subject. Don't you get your bar scores
soon?" Rena asked.

Hollis nodded. "Not soon enough. That's all I need, to
bomb there too." She sighed. "You guys, I've got to go home.
Stephanie still can't get around that well."

Rena shot a look to Mark.

"Okay, we'll take you home," Mark said. "But you should
know that it was Stephanie's idea to take you out to dinner. She
said you're getting on her nerves."

HOLLIS GAVE A gentle tap on the bedroom door.

"Who is it?"

Pushing the door open, Hollis walked into the room and
sat on the end of the bed. "Very funny. How are you feeling?
What's all this?"

"I couldn't be better." Stephanie peeked over the rim of her glasses and pointed to a stack of pages beside her laptop. "This, Ms. Morgan, is the stash you took from Cathy's apartment. You know, the window sill safe."

"I don't understand. How did you ....? And you are doing what with it?"

"I didn't mean to pry but you left it on the coffee table." Stephanie smiled. "What can I say? I'm a forensic technician; there's nothing for me to do all day and I'm nosy."

Hollis picked up the top article. "I wasn't able to make the Fields connection, so I think this was going to be her next story."

"Then that makes sense. Most of this stuff is about *Transformation* exposés. I did a little research of my own; they were all later recanted or settled. I couldn't find any that linked to Fields' non-profits."

"I know. I checked for them too."

"I always knew they had to be getting sued for all the crazy stuff they printed." Stephanie pushed a paper toward Hollis. "So I did Internet searches and found out what I could about the paid-out dollars, but *Transformation* attorneys are smart enough to ask for a settlement so they don't go to court, and they don't have to tell you how much it cost them."

"I know. I got to experience that moment."

"I don't know how they can keep it up. Their subscriptions can't bring in that much."

Hollis dug into her tote and tossed *Transformation*'s annual report on the bed. "My thinking exactly. What do you think is going on?"

"Well, according to court documents, so far *Transformation* offered a total of six million last year in settlement claims, and I found another two million this year. They must be self-insured; I can't imagine an insurance institution would take on that claims record."

Hollis frowned. "Those numbers are confidential. How did

she, or you, find out the payments?"

Reaching over to her nightstand, Stephanie picked up a tissue and wiped the lenses of her glasses. "Like I said, I'm a forensic tech. I tracked the original suit number to the closed court file. I'm assuming she did the same."

Hollis squeezed her eyes shut; the day was moving into overload.

"Wait, are you sure Cathy knew about the claims?"

"See these sketches? They're of CEOs, financiers, and politicians. It took me a minute—well, a long minute—to figure it out, but these are people *Transformation* had written about over the past five years, and the magazine had publically promised to delve even deeper for a 'real scoop.' " Stephanie put up her fingers in mock quotes. "But instead they settled."

"But Cathy didn't have access to closed court cases." Hollis sat straight up. "Wait, I know. She probably sneaked into *Transformation*'s accounting office." She glanced down the sheet. "Hey, I know this lady. She's on one of those receipts in Cathy's stack."

Hollis' mind was doing somersaults as another picture was forming—a picture that made much more sense and could be proved. "Stephanie, you are fantastic."

Stephanie shrugged and then winced. "Ouch."

Hollis stood. "I've neglected my nurse duties. Are you hungry? Need a pain pill?"

"No, Nurse Betty, I'm fine." Stephanie laughed. "Actually, when I go for my checkup next week I'm going to ask the doctor for a return-to-work slip."

Hollis leaned against the door jamb. "And I need to return home."

"Don't. Having you around is helping me heal."

"I hope so. But I need to deal with my own life issues." Hollis shook her head. "I'm the reason you're in this position in the first place. You took that bullet for me."

Stephanie feigned a look of dismay. "Rats, I've always wanted to say that. Look, stay here at least until I get my release from the doctor," she said. "You wouldn't abandon an invalid, would you?"

Hollis smiled. "I guess I—"

"Good. Oh, I almost forgot to tell you. Your detective called. I told him you were out to dinner. He said he would call back."

Hollis felt herself blush. "He's not *my* detective."

"Sure, whatever." Wearing a knowing grin, Stephanie lay back down on the pillow and almost immediately fell asleep.

Hollis straightened the room. Then she wrote a brief note indicating she was running errands and would return in an hour. After dashing home, she gathered the mail and enough clothes for another couple of days. Other than bills and a couple of ads there was nothing important.

Still, it felt good to be back. Hollis hadn't expected to feel apprehensive about returning home after the shooting. She refused to feel like a victim, and she had missed her own nest. She could help Stephanie just as well by dropping in on her.

She went through the condo, making sure everything was locked up. But she could not help but notice that it was time to clean out her refrigerator. She liked a clean home. Her condo needed a major going over, but over the past few months, her studies had supplanted any free time. As always, her best thinking happened while she was doing mundane tasks. Throwing away mystery leftovers was as far away from *Transformation* as she could imagine. She emptied the vegetable drawer, which had formed the beginnings of a mold colony.

In the dining room she paused. The table gleamed with filtered light from the shutters. Maybe it would be nice to fix dinner for a few friends. She could keep it real simple. They wouldn't have to stay long.

A slow smile crept on to her face. She finally had a plan, and that always put her in a good frame of mind—a plan to nail a killer.

CHAPTER TWENTY-FIVE

~~~

MORTON PHOTOGRAPHY OPENED at ten a.m. on Wednesdays. Hollis intended to be there shortly thereafter. This time, if she were right, Joe Morton would identify Cathy's and Gail's murderer. She wondered briefly if she should contact Cavanaugh with her suspicions, but just as quickly she discarded the idea. He wouldn't appreciate how she got the dots to connect. Afterward, if she found the proof, he would be the first to know. She'd learned her lesson about interfering with police work.

When Hollis left Stephanie that morning, she was happily talking to her family on Skype. She had to make a quick stop at her condo before going to see Joe; she had left her cellphone there the evening before. It was definitely time to return home.

The doorbell rang.

It was her neighbor, dressed in cycling gear, helmet in hand. Hollis thought her name was Donna, or Lana? Something ending in an "uh." Maybe Tanya?

"Hi, I'm Christy. I live across the street. I accidently got your mail, but I've been out of town for the past two weeks and didn't pick up my mail until today. Hopefully it's not anything important."

The young woman handed her two envelopes and a postal attempted-delivery slip. Hollis glanced through the material quickly.

"Thank you. It's just routine bills." She raised the slip. "And I guess I need to go to the post office."

"Sorry." Christy waved goodbye and took off at a modest trot.

Hollis called out a thank you and closed the door.

The post office was not far from the photography studio. Hollis glanced at the clock. First, the quick stop at her condo to pick up her cellphone. After seeing Joe, she'd stop by the post office. If he could identify the pictures, she wouldn't have to depend on the thumb drive.

JOE WAS LATE opening the studio.

Hollis kept a lid on her irritation while she waited, gazing out the car window to pass the time. It was one of her favorite weather days—foggy, cool and silent. She had been so intent on solving Cathy's clues, she had almost neglected to notice her favorite weather. She tilted her head back and closed her eyes. A noise caught her attention; Joe was opening the studio door.

He turned on the lights as she entered, the bells tinkling over her head.

"Hey, what's up?" he called from behind the counter.

"Hi Joe, I have those pictures for you to look at," Hollis said. "I'd like to see if you can recognize them as the men you saw that day in front of the Remington Building."

He beckoned for her to sit. "Sure, give me a minute. I'm on my own. Amber isn't coming in today."

He opened blinds and turned on lights. Finally he sat next to Hollis on a counter stool.

"Okay, let's have a look." He pushed aside a stack of digital card readers spread out on a work table behind the counter and turned on a desk lamp.

Hollis pushed a folder and let its contents slide out onto the table.

There were three pictures, two from magazine articles and a third from an annual report.

He examined one. "I don't know, maybe." Joe took a closer look then put it aside.

"Just take your time." Realizing that she was holding her breath, she took the one back.

He picked up the second article. "Yeah, now *this* guy looks familiar. It's been a while, but I think he's one of them. Wait." Joe slipped on his eye glasses. "Yeah, I remember seeing them together on those steps. Let me see the first one again."

Hollis passed it back.

Joe smiled with accomplishment. "Until I saw the second guy, the first one didn't catch my attention."

Hollis looked down at the picture and was ready to leap for joy. "Joe, I think I love you."

He turned a warm red. "I think you need to get to know me first." Joe lifted his glasses. "Let me see the book."

"It's an annual report." She turned to the page where she had placed the Post-it. "Was he the third man?"

Joe Morton squinted at the picture. "No, no he wasn't one of the three."

The disappointment hit Hollis hard. She hadn't realized how much she was counting on Joe's ID.

"He came later."

Her head snapped up from the report. "Later?"

"Yeah, I told you that Cathy wanted me to take pictures of everyone who came out of the building over a half-hour period. Well, this dude came out about ten minutes after the other three had left." Joe took off his glasses. "I remember, 'cause he's a sharp dresser. I recognize the suspenders. He had them on that day, too."

Harold Roemer, Arlo Mueller and—surprise of surprises—Mr. Suspenders, Carl Devi.

"Thanks, Joe. You have made my life a whole lot easier."

"Good, I hope I've helped you find Cathy's killer," Joe said. "Er … maybe you can come by sometime when you're in the neighborhood and just say hello." He smiled shyly.

Hollis smiled back. "I'll do that."

THE POSTAL CLERK handed her a small brown padded envelope with no return address and a noticeable lump in the middle. Hollis' heart started to pound. In the parking lot, she tore into the package. Her hopes were confirmed.

Yes! It was Gail's thumb drive.

She was still waiting for her home insurance to replace her laptop, so she would have to use her firm's computer. She almost ran to her office.

"Hey, Hollis, you too, huh?" One of the associate attorneys was leaving the coffee room when she entered. "All the big guys are in a management briefing. The associates are in computer training. So far there are just us paralegals here today."

Rushing toward the hallway, Hollis called over her shoulder, "Misery loves company."

She quickly booted up her computer and inserted the thumb drive. In minutes she was able to view on the screen what appeared to be not an article, but detailed notes. She hit the print key.

Out came twenty-three pages of notes, comments, and quotes. There was even a small spreadsheet calendar.

Finally.

CHAPTER TWENTY-SIX

H OLLIS TRIED TO recall when she first had the idea to break into *Transformation*'s business files. She probably should have done it in the beginning. She made a quick stop at police headquarters and dropped off the thumb drive for Cavanaugh. She felt a little guilty, because she hadn't told him to expect it. It might be a couple of days before he knew she had. She couldn't deny that she wanted to be the one who discovered Cathy's killer.

Now, standing in the elevator, she was more than ready to get the proof she needed. The doors opened and she faced her first hurdle.

"Hi, Phillip." She walked quickly past the reception desk. "I have some accounting files to return to Mr. Devi. I'll just take them to the file room."

She had interrupted his texting. Phillip frowned. "Uh, it's late. I don't think—"

"It's okay, I know where it is."

"But Mr. Devi's out of the office."

Hollis nodded. "I know."

She was counting on it.

The file room was located next to the stairwell. Hollis actually did know where she was going. Once she had helped one of the secretaries with carrying a load of background information from the library, but it was the Accounting Section and not the file room where she was headed.

It took a few minutes to figure out how the files were categorized, but she knew she was on the right track when she finally spotted Harold Roemer's file. But her moment of accomplishment was dashed when it only contained the same article she found in Cathy's condo. Her fingers moved down the row to Arlo Mueller. His file was empty.

She hit her fist against the flat panel at the end of the shelf. If *Transformation*'s files had been purged, she was going to have a difficult time convincing Cavanaugh of her hunch. And she had to admit all she had was a hunch.

She put everything back as she found it.

Hollis gave Phillip a quick wave goodbye. Absorbed in a phone conversation, he barely nodded at her.

On the elevator ride to the parking garage she gathered her thoughts. *Transformation* had run stories about Roemer's extortion for weeks. But nothing on Mueller, even though he was smack in the middle of the same scandal. His empty file didn't make any sense, except if you read Cathy's notes.

It had taken her a long time to figure out what Cathy's files and notes really meant. Hollis realized that Cathy may not have figured it out for herself. It was unfortunate she hadn't seen the true story before she wrote the article. She had been given the Fields' story as a red herring—a decoy from the truth. Cathy had been getting close—but not close enough—when she was murdered.

The elevator doors opened. A flash of red distracted her so that when the blow came to the side of her head, the pain was sharp and deep.

And then she was falling.

***

THE METAL FLOOR was cold and hard.

Hollis tried to open her eyes to see in the darkness, and then she realized that they *were* open. It was pitch black. She was no longer in the garage. Raising herself on one arm caused her head to swim. She waited for her head to settle and then sat up.

"Good, you're awake." The voice seemed to come from above and behind her.

She tried to turn around, but her vertigo wouldn't allow it. She waited and then looked over her shoulder. The metal sides and floor reminded her of a trash dumpster. Her eyes were adjusting.

"Where am I?"

"Where no one will ever find you, Miss Morgan."

The voice came from what appeared to be some kind of ledge or ramp, high above her. It sounded like an echo on a speaker phone that bounced off the container walls, a container that looked to be in a basement.

Hollis stretched her legs and tried to stand. The container rocked, throwing her off balance. She fell hard against the floor.

She shrieked.

"No, no! That's not going to happen. There is no escape. You're going to have to die. I tried to warn you off, but you wouldn't listen."

She tried to ignore the taste of fear that was creeping up into her throat.

"You can let me go. I didn't see you. I don't know what you look like. I don't know who you are. Just go. You can leave here. I'll count to one thousand and you can just go your own way."

Hollis thought she heard him chuckle. She was pretty sure the voice was a male.

"I'll give you two choices. I will bury you where you sit and they will find your bones years from now. Or, if you cooperate, I'll bury you where they'll find your body in a month or so. Either way you're dead, but under the second option your friends and family won't worry or stress for years."

She tried to hold back the tears, but his words were shaking her resolve.

"Why me?"

"You, as they say, know too much."

Hollis mentally forced her heart to beat normally and her breathing to slow. She had to buy time. Someone would start to look for her. Although at the moment she couldn't think of any reason they would. She had checked out of the office, and Stephanie, who had gone back to work today, would probably be home resting.

No one would miss her.

"What is it that I know?"

"We're not having a conversation, and I don't have all evening. What's it to be?"

Hollis took another deep breath. "I'll cooperate."

She could hear scrambling down stairs and a scraping of metal on metal. In the corner of the container a small door slid open.

"Crawl out backwards. I have a gun." The voice, now familiar, sounded indifferent, as if ordering fast food.

In the deep darkness of the basement, her knees scraped against a concrete floor littered with debris. She blinked rapidly, trying to get her eyes to adjust to the new shades of dark. Now clear of her metal box, she still had her back to her tormentor.

"You might as well let me face you," she said. "You let me hear your voice." She braced and pulled herself up against what was her first guess—a dumpster.

"Sure, turn around and walk to that car at the end of the ramp."

Hollis, still a little dizzy, turned to face Carl Devi. He looked grim.

"I can't—"

"The car's over there. Move."

She had to stall. Hollis looked around with a deliberate

slowness to find anything that would increase her odds. She stumbled.

"Wait, I have to—"

This time the blow slammed into her forehead, and it was only a short distance to the floor.

WHEN SHE CAME to, she was in the trunk of a car.

It wasn't moving. There was only silence. Hollis' hands and feet were wrapped in duct tape. Her mouth was taped and her head throbbed horribly. This time she couldn't stop the tears.

She had to get out. She kicked at the trunk lid, but she couldn't get leverage. She wasn't strong enough. Devi must not be around. He had left her here. Was this where she was going to be found in a couple of months? Or was he coming back to drive the car into the bay? How long had she been here? If she could turn around onto her back, her legs would have more thrust. She would kick out the rear light.

"Hollis?"

She stopped squirming.

The male voice was a little above a whisper. "Hollis, are you in there?"

She tried to call out, but her mumbled words fell back into her mouth. She thrust her legs against the trunk roof.

"Okay, I can hear you. I have to break into the car. It's locked with an alarm system." He moved away. "I'll be back."

Vince?

Hollis was frantic. She didn't want him to leave. Supposing Devi came back? She tried to scream "no," but she was met with silence again.

After only a few minutes—which seemed like an hour—he was back.

"Hollis, I'm here. Move away from the edge of the trunk. I'm going to knock out the light and try to open the trunk from the inside with a metal bar."

She did as she was told.

With a loud crack, glass splattered into the trunk. She could see a lighter shade of darkness. The metal bar shoved back and forth through the opening, scraping the interior mechanism. Minutes passed. How long would Devi be away?

"Hollis, can you see me?"

Her response was muffled.

"That sounds like a yes. Look, we don't have much time. I have to poke around inside to see if I can find the safety lock to pop the trunk. I don't want to hurt you. Move back as far as you can."

Hollis managed to scoot her legs a few inches.

"I can see you, Hollis; I can see your legs." Vince talked while he struggled to angle the bar. "We're going to get you out of this. I saw this in a movie."

She was glad he couldn't hear her moan.

How long did they have before Devi returned? It all seemed like a nightmare. Vince was talking more to himself than to her. The incessant scraping of the bar seeking a possible latch was grating on her nerves.

This must be what it's like to go crazy.

Then the lid popped open.

"Hollis, I did it," he yelled and put down the bar. "Here, let me take off the tape."

She endured the pain of the slow pulling. It felt like she lost the first two layers of her skin.

Vince looked at her with dismay. "Don't cry. I'm sorry."

It was done.

She tried to lick her lips but she had no saliva.

She found a small voice. "Thank you."

Vince was already starting on her legs.

"Good thing you wore pants. This won't hurt nearly as much." He ripped off the binding.

Hollis flexed her legs and scrambled out of the trunk.

"Hurry, Vince, we've got to get out of here." She turned to let him pull the tape off her arms. "Just yank it off. It'll be over sooner."

In a few seconds her arms were free.

"What I wouldn't give for some water." She stood in the middle of what looked like the basement. "What time is it? Where are we?" Her voice cracked. "How did you get in here?"

"I followed you from the garage maybe a couple of hours ago." He looked down. "I know you tol' me not to follow you but I saw him hit you—"

Hollis held his shoulder. "Vince, forget that. Where are we?"

"It's like some kinda second floor basement. There's a car elevator that goes to this floor. We're a floor below the main garage. I followed the way he came in and waited." He paused. "I saw the gun."

She realized he was a scared kid and trying hard not to show it. Her head was throbbing again.

"Do you have a cellphone?"

He shook his head.

"Show me the elevator."

She wobbled as she followed behind him. They had to move faster.

They were halfway across the floor when the sound of a cushion of air whooshed from the oversized elevator and the doors rumbled open.

Carl Devi's face blanched from surprise to anger in seconds. Hollis faced him. She could feel Vince's presence behind her.

"Well, Ms. Morgan, if it wasn't for your current circumstances, I'd say you lived a charmed life," he said. "Who is this piece of waste?"

She tried to lick her lips. "Why did you kill Cathy, Carl? She never knew you were behind the fake Fields story."

She shifted her weight on her weakened legs.

He snorted. "So you figured out it was me. You're a very smart lady."

Hollis tried to formulate a plan as she kept him talking. "No, I'm not that smart. I can just smell lies. It had to be you. Cathy's story was a fraud. You fed her bad data, bad leads, and sent her to bad community groups. You even let the story leak to Bartlett ahead of time. You knew Dorian Fields couldn't allow lies to get published. He would have to sue and that would stop Cathy."

Devi glared at her.

She sensed rather than saw Vince's slow movement shift of just a few inches.

Hollis continued. "You needed to distract her from the other story she was starting to get into. A story that would end *Transformation*—and you—for good, a story about how you were blackmailing your subjects under investigation. In exchange for not being named in your trashy tabloid, *Transformation* would always offer to settle for one amount, and you would get a kick-back. But where did you come up with all the money? How did you get insurance?"

"You *are* smart." He looked behind her. "Don't even think about it, kid." He motioned for Vince to back away. Vince stopped with both hands behind his back.

Hollis noticed the faint gleam of the bar he was still holding.

"All that money, where did it come from? So what happened, Carl? Did you get sloppy?"

"It was only a matter of time before she would stumble onto my side business. There was too much at stake."

Hollis worked on keeping the fear out of her voice. They didn't have a chance. She could talk Devi blue in the face and he would still be blocking the elevator with a gun in his hand.

"Don't think I don't know you're trying to stall." He smiled. "It won't work. I'm just waiting for the building to empty."

Hollis felt her anger build and adrenaline pump to her heart. It made her sore head a little dizzy, but her peripheral vision, now sharp, sought the figure that was inching toward Devi.

"And Gail?"

"Gail? She would be considered a nuisance on a good day. Nobody will miss her." His voice was ominously quiet. "I had to kill her. She might have started putting things together about a few conversations I had with her supervisor and come up with me."

Without taking his eyes off Hollis, he reached behind him and turned on an overhead light.

"Now I'll have to do it again." He pointed toward Vince. "And again."

Hollis blinked from the glare and licked her lips. "I'm not stupid. I told Detective Faber to meet me here."

"Really, how did you call him?"

"I have a cellphone. I placed the call just before I got here."

Devi shook his head. "No, Hollis, you didn't. I put a bug in your phone that day your condo was broken into. *Transformation* prides itself on its spy toys. I've been tracking your calls for days."

Hollis thought back to all the visits to centers that knew ahead of time she was coming. She needed time. "Is that gun really loaded?"

"As a matter of fact it is." He laughed.

"You know what? I hate threats. Did you know I've been to prison?" She smiled. "No, I can tell you didn't. Well, believe me, you won't last for a minute in the slammer." She nodded toward Vince. "He's just a kid. Hold me hostage. Let him go."

Devi looked over at Vince, who still stood with his hands behind his back.

"You've been awfully quiet, kid." Devi motioned with the gun. "Turn around and let me see your hands."

Hollis sighed as Vince held out the metal bar.

"Drop it and kick it over here."

Vince said nothing as the metal clanged loudly against the floor. Hollis felt his dejection as they both saw their last hope slide away.

"I'm not ready to let go of my lifestyle just yet." Devi moved toward her with deliberate slow steps. "Once you've killed two people, a third and a fourth is just an inconvenience."

He looked down at his watch. "Only a few minutes more to be safe."

Hollis could feel her forced bravado fading. "Then what?"

"Then you and your friend will get a little tap on the head and go back into the trunk for a final ride."

Hollis almost shut her eyes, but just then she saw Vince yell and lunge toward Devi. Devi was caught off guard but had enough time to push Vince off him. Hollis picked up the metal bar as she ran toward the struggling men. Vince and Devi wrestled, but Vince didn't have the strength to withstand Devi's bulk.

Hollis stood to the side holding the bar in her hand like a bat. She knew she would have only one chance to get it right. She jumped when seconds later a shot rang out and Vince fell to the ground. Devi stood over him, breathing with deep gasps.

Hollis moved quickly. Without hesitation, she swung the bar as hard as she could. Devi fell at her feet, blood gushing from his skull.

"See how you like getting hit in the head."

Her chest heaved as she tried to catch her breath, but for only a moment. She ran to Vince. Blood was already pooling under him. She turned him over and held him close. His thin body had only a faint pulse. She had to get help. She rose from the floor when his eyelids fluttered. He looked at her without speaking, and then his eyes closed.

"No!" Hollis screamed.

CHAPTER TWENTY-SEVEN

───ೲ───

THE ELEVATOR DOORS opened onto the lower garage, just as Vince had described. Hollis ran to the passenger elevator and began pushing buttons, leaving smears of blood on the panel. But she could go no farther. She needed a security card to call the elevator. Tears blurred her vision; she wiped them away as she ran to the ramp. There was the wrought-iron gate, but it needed a remote to open.

"Help!" she screamed through the bars. "Help me!"

After what felt like an hour of constant screaming, she heard footsteps. She tasted her own blood. Her throat must be bleeding.

"Hey lady, this is security. We're on our way. You're going to be okay."

Shaking with sobs, she slid down to the ground.

SHE SAT OFF to the side as the police and medics rushed past her to the elevator.

It was over.

She put her head in her hands. She felt arms around her shoulders.

"Hollis, it's John. Are you okay?"

She looked up and grabbed him by the arm. Her voice rasped. "John, help Vince. He's down there with Devi. He's just a kid. I think he's …. He might be …."

"Shh, stay here." He patted her hand. "I'll be back. I'm going down."

The elevator rumbled its return.

An EMT carrying a bag knelt beside her. "Lady, let me look at your head. You had a couple of real nasty blows."

As he examined her, she winced with pain.

He reached into his bag. "They're more bloody than serious; fortunately they missed your eye." He quickly cleaned and swabbed her head and hair with a wet cloth then wrapped both in a gauze bandage. "You need to get to the hospital, too. You're going to need an MRI and possibly an EEG."

The elevator doors opened again, and two stretchers rolled out, one with an IV setup, rushing toward the ramp. The other held Devi.

Devi was handcuffed to the bed, his head swathed in white bandages. His stretcher was followed by Cavanaugh, who spoke rapidly with Faber as he strode up the ramp.

Hollis hobbled over to the ambulance, leaning on the arm of a medic. She looked down at Vince. "Is he … is he alive?"

The medic looked at her sympathetically. "He's likely concussed. We need to get him to a hospital as soon as possible. He took a bullet to the chest. He's going to need surgery."

She picked up his pale hand with its badly bitten dirty nails. "Oh, Vince, why did you do it?" Her tears fell on his cheeks.

His lashes flickered and his eyelids fluttered open halfway. She blinked because she thought she saw the vaguest hint of a smile. She put her ear to his lips and heard him whisper, "Because I could."

CHAPTER TWENTY-EIGHT

"**W**HEN DID YOU figure out the motive?" Faber stood next to Hollis' bed holding her hand while they waited in the hospital emergency room for her MRI results.

She knew he was trying to distract her from worrying about Vince's surgery. Stephanie stood on the other side of the bed looking equally concerned. The edge of her own bandage was still evident under her sweater.

Hollis' forehead wrinkled. "Actually, it was another matter that had my attention—a probate case about missing heirs." She took another sip of water. "One of the principals hid his brother's suicide for over thirty years so he could share his own anguish with my client, who was already guilt-ridden. That's when it clicked in."

"I must be tired." Stephanie brought a chair close to the bed and sat. "How did you connect his guilt to this?"

"It kept bothering me that Cathy could have been so sloppy with her due diligence. It wasn't like her. Then we kept coming up with articles and pictures that had nothing to do with Fields. Then there were the interviews at the charities that Devi was feeding us .... Everything pointed to libel," Hollis paused with a wince, "but it wasn't until I showed Cathy's photographer friend Devi's picture, and he identified Devi as being one of three men he saw in a meeting. That turned our assumptions upside down. I did a little extra digging. It was really you, Stephanie, who put me on the right road. Remember with the settlement dollars?"

Hollis continued, "Devi was blackmailing Arlo Mueller. His wife may or may not know. Devi discovered that Mueller was involved with that Roemer scandal and offered to keep Mueller's role quiet if he settled with *Transformation* and gave Devi a kickback."

Faber nodded. "I get it. Devi needed to deflect attention because Cathy was hot on his trail. He gave her a fake Fields of Giving story and kept up the pretense."

Stephanie frowned. "Okay, I'm with you there, but what had this to do with the probate case?"

"Eric Ferris, who did time for his brother's murder, learned later that his brother had actually committed suicide. Realizing our client was full of guilt, he steamed open her letters to him, but he didn't want her to know he'd read them. He would steam them open, read them, and send them back—'return to sender.' He hid the real story even from his family." Hollis paused.

Hollis felt her energy waning. "There was never a Fields story. Devi had set everything up to protect the real story, and himself. And I bet it wasn't the first time."

The doctor entered.

"Hollis Morgan?" He came over to the bed. "Your brother is conscious. He wanted me to let you know he's okay. He's groggy and won't be able to have visitors for another day or so, but with proper care he's going to be fine."

She nodded and stopped Stephanie's protest with a look. John pretended to be engrossed in the curtains.

Hollis smiled. "Thank you, doctor. Please tell my brother when he wakes up that I have his GED application."

EPILOGUE

~~~

S HE HESITATED TO look at the CalBar website for her scores. Much like parents who want to wait to know the sex of their unborn child, Hollis was torn between wanting to know and remaining in the dark.

"Give me your username. I'll check for your scores." John Faber reached for the laptop sitting on the kitchen table.

"Yes, wait … no." She pushed his hand away.

"Hollis, we've been sitting here staring at the computer for twenty minutes," he said patiently. "We're going to be late for our dinner reservations. You don't even know if the scores are posted."

"The scores are in. There's a notice on the CalBar website saying six o'clock on Friday."

John looked at the clock on the oven. "It's almost seven."

She sighed.

He held her hand. "Okay, tell me how it works."

"I go to the results page and type in my applicant number, and then type in my California State Bar file number. If my name pops up on the screen, it means I passed. If my name doesn't pop up, the screen says the website didn't recognize my information for any number of reasons. One being, I didn't pass."

"No matter how long you wait, whether you passed or not, it's already decided, right?" John said. "You want me to go to the website?"

"What if it's bad news? I've worked for this for so long." Hollis sighed. "I don't think I'd be able to eat dinner either way."

He took her by her shoulders and turned her to face him. "During these past weeks and months, all you've done is face the truth. Compared to what you've been through, this notice is nothing."

Hollis nodded. Enough was enough. John was right. She quickly tapped in her ID. Squinting with one eye, she read the simple sentence:

"The Committee of Bar Examiners is pleased to inform you that you passed the California Bar Examination."

R. Franklin James grew up in the San Francisco East Bay Area and graduated from the University of California at Berkeley. She and her husband currently live in northern California.

Sticks & Stones is the second novel in the Hollis Morgan Mystery series, following *The Fallen Angels Book Club*.

You can find R. Franklin on the Web at:
www.rfranklinjames.com.